Mark Anthony Benvenuto, Heinz Plaumann (Eds.)
Green Chemistry in Government and Industry
Green Chemical Processing

Green Chemical Processing

Edited by
Mark Anthony Benvenuto

Volume 5

Green Chemistry in Government and Industry

Edited by
Mark Anthony Benvenuto
and Heinz Plaumann

DE GRUYTER

Editors
Prof. Dr. Mark Anthony Benvenuto
University of Detroit Mercy
Department of Chemistry & Biochemistry
4001 W. McNichols Rd.
Detroit, MI 48221-3038
USA
benvenma@udmercy.edu

Dr. Heinz Plaumann
Adjunct Professor - Chemical Engineering
Wayne State University
5050 Wayne Anthony Drive,
Detroit MI 48202
USA
Cj1709@wayne.edu
734-747-4770

ISBN 978-3-11-059728-8
e-ISBN (PDF) 978-3-11-059778-3
e-ISBN (EPUB) 978-3-11-059805-6

Library of Congress Control Number: 2020940944

Bibliographic information published by the Deutsche Nationalbibliothek
The Deutsche Nationalbibliothek lists this publication in the Deutsche Nationalbibliografie; detailed bibliographic data are available on the Internet at http://dnb.dnb.de.

Cover image: scyther5/iStock/Getty Images Plus
Typesetting: Integra Software Services Pvt. Ltd.
Printing and binding: CPI books GmbH, Leck

www.degruyter.com

About the series

Green Chemical Processing is a continuing series of volumes composed of refereed chapters, with upcoming volumes having submission dates of 15 June and 15 December each year. All areas of green chemistry, pending as well as established, are considered and welcome.

If you are interested in contributing a chapter, please contact series editor Mark Benvenuto, of the University of Detroit Mercy, at: benvenma@udmercy.edu concerning the appropriateness of your topic. We are interested in any and all new ideas that examine any of the twelve principles of green chemistry.

For more information on all previous and upcoming volumes of *Green Chemical Processing*, see
https://www.degruyter.com/view/serial/GRCP-B

https://doi.org/10.1515/9783110597783-202

Contents

List of contributing authors

Mark Anthony Benvenuto
University of Detroit Mercy
Department of Chemistry & Biochemistry
Detroit, Michigan
United States
benvenma@udmercy.edu

Christina Britton
Meijer Inc Ringgold standard institution -
Properties and Real Estate
Grand Rapids, Michigan
United States
Christina.britton@meijer.com

Alice Delia
Prism Analytical Technologies
Mount Pleasant, Michigan
United States
a.delia@pati-air.com

Joseph C. Furgal
Bowling Green State University - Chemistry
and Photochemical Sciences
Bowling Green, Ohio
United States
furgalj@bgsu.edu

Susan Kozora
International Automotive Components -
Advanced Development and Material
Engineering
Southfield, Michigan
United States
skozora@iacgroup.com

Dale J. LeCaptain
Central Michigan University - Chemistry &
Biochemistry
Mount Pleasant, Michigan
United States
ecap1dj@cmich.edu

Chamika Lenora
Bowling Green State University - Chemistry
and Photochemical Sciences
Bowling Green, Ohio
United States
rlenora@bgsu.edu

Mary Katie Martin
Central Michigan University
Chemistry & Biochemistry
Mount Pleasant, Michigan
United States
kmartin@ucdavis.edu

Attila E. Pavlath
Past President, American Chemical Society
Albany, California
United States
attilapavlath@yahoo.com

Erik A. Petrovskis
Meijer Inc Ringgold standard institution -
Properties and Real Estate
Grand Rapids, Michigan
United States
eapetrovskis@gmail.com

Heinz Plaumann
Adjunct Professor - Chemical Engineering
Wayne State University
Detroit, Michigan
United States
Cj1709@wayne.edu

Rose Ann Ryntz
Ryntz & Associates, LLC - President
The Villages, Florida
United States
roseryntz@comcast.net

Susan C. Walkinshaw
NSF International
Ann Arbor, Michigan
United States
walkinshaw@nsf.org

Roby Wong
Qualitative Green Consulting, LLC -
Environmental Consulting Sector
Hattiesburg, Mississippi
United States
roby.wong@qgconsulting.com

Heinz Plaumann and Mark A. Benvenuto

1 It's not easy becoming Green ... but we're getting there!

Abstract: This volume contains contributions from a very diverse group, starting with a former president of American Chemical Society, and with members in industries including a leading grocery chain in the United States mid-west, technical experts in the transportation industry, and including academia and government agencies. We look forward to the growing momentum in greening our world with ongoing progress in this growing and evolving series.

Keywords: green chemistry in industry, plastics, green chemistry in automotive, chemistry and lumber

1.1 Introduction

"It's not easy being Green" proclaimed Muppet superstar Kermit the Frog in the iconic song, but this volume teaches that we are indeed getting better at it. What might be called "The Parade" is growing and gaining momentum. Figure 1.1 and Figure 1.2 show examples of efforts designed to make our society greener, and at the same time raise awareness of the issue.

This fifth volume again includes contributions from a distinguished list of authors, covering a diverse range of topics. Any reader with a curious bent will be cerebrally stimulated by the topics, their relevance, and the range of writing styles. Many Green Chemistry thrusts have been directed in three broad areas: transportation, energy sources, and the problem of disposal. These might now be considered the usual "3 Rs" – Reduce, Reuse, Recycle. With the following brief chapter summaries, let us whet your appetite to consume and then digest what we hope can be called this delightful volume.

1.2 This volume

The chapters in this volume will indeed enlighten the reader to how broadly the principles of green chemistry are being applied. The greening of our chemistry, indeed

This article has previously been published in the journal Physical Sciences Reviews. Please cite as: Benvenuto, M. A. It's not easy becoming Green ... but we're getting there! *Physical Sciences Reviews* [Online] 2020, 5. DOI: 10.1515/psr-2019-0019.

https://doi.org/10.1515/9783110597783-001

Figure 1.1: Garbage receptacle at a university proclaims the green chemistry theme.

our science and our world is necessary. Enjoy these many diverse examples as "role models", challenges, and encouragement.

1.2.1 Biodegradable polymers – why, what, how?

We are honored to have Dr Attila Pavlath as the distinguished author of this volume's lead chapter. A former president of the American Chemical Society, his career has included a broad range of contributions to the chemical enterprise – including the holder of many patents, winner of many awards, author of numerous technical articles and books, long-time professor, now emeritus. Many have said the curious mind, needs to know, and Dr Pavlath's life's passion has been the pursuit of knowledge to share with Humanity. His timely chapter here is of considerable import.

Polymers are ubiquitous and have contributed much to our current quality of life, be it in areas such as transportation, energy, medicine, recreation, construction. In truth, there is no current human endeavor untouched by some polymeric influence.

Figure 1.2: A U.S. beauty salon chain markets their support to environmental concerns.

In this excellent expose “Biodegradable Polymers: Why, What, How?” Dr Pavlath walks us through from the synthesis craze in the twentieth century to tailor-making polymers to meet specific properties, and how we are now burdened somewhat with the backdrop of, “from cradle . . . to now what?”. Clearly, a strong and obvious focus is the trend to design biodegradable polymers to be “yummy bug-food” to facilitate their decomposition. Many examples and references make this a worthwhile read as well as a reference source.

1.2.2 Sustainable materials in automotive

Dr Rose Ann Ryntz holds a Ph.D. from the University of Detroit and an M.B.A. from the University of Michigan. Her academic qualifications however pale in contrast with being a true cornerstone in the transportation industry where she has contributed to the commercial success of several companies. She has received many awards from such groups as the Society of Plastics Engineers, the

American Chemical Society, and the Engineering Society of Detroit, to mention only a few. She and her co-author Susan Kozora are with the International Automotive Components Group.

The contribution of Dr Ryntz to this volume reviews the breadth of polymeric materials in this industry, with introduction to specific materials, the properties which render them useful, and their ultimate fate – be it recycling or repurposing. They also include a superb section on legislative issues along with sustainability issues in the industry. An excellent section about bio-based versus recycled materials is truly thought provoking.

1.2.3 Meijer and carbon management

Dr Erik Petrovskis holds a Ph.D. in Environmental Engineering from the University of Michigan and has been a lecturer there. He has over twenty years of experience as an environmental consultant and serves as Director of Environmental Compliance and Sustainability for Meijer. His co-author, Christina Britton, holds a MS in Environmental Policy from Bard College and is a Sustainability Analyst for Meijer.

Meijer is a family-owned retailer in the Midwestern United States. The 84-year-old company has always lived by its commitment to environmental concern. It currently operates over 240 stores in six states with its concomitant carbon-footprint from heating, refrigeration, lighting and supply chain/logistics. Erik and Christina present these challenges and how Meijer has dealt with them in exemplary fashion – an excellent model for other industries.

1.2.4 Applying green chemistry approaches to EPA-standard method of analysis of dioxins

Susan Walkinshaw is an analytical expert with the EPA, and her chapter clearly and eloquently speaks to this, comparing a number of different extractions methods to recover specific target molecules for analysis. The journey begins with a problem statement, selection of method, sample preparation and solvent considerations. The analysis includes discussion of the hazards of the various solvent options along with their respective costs.

1.2.5 Green routes to silicon-based materials and their environmental implications

Dr Joseph Furgal received his Ph.D. in Materials Chemistry at the University of Michigan and is now an assistant professor at Bowling Green University. His co-author Dr Chambika Lenora holds a Ph.D. in inorganic chemistry from Wayne State University.

They start their contribution by indicating that "silicon based materials have their origins in the Stone Age" with quartz and other rocks to make the first tools. Their historical treatment covers the first reported interactions of silicon and carbon in the late 1800s leading to the development of natural rubber replacements with silicone rubbers during the 1930s and 1940s.

Silicon-based materials play very important roles in numerous industries, but especially in the computer and related chip industries. A significant contributor to "greening" this industry has been the conversion of the bio-based raw material (specifically rice hull ash) into useful silicon building blocks rather than the low reactivity mined silica sources.

1.2.6 Greening within the lumber industry

The lumber industry is among those cornerstone industries which have not been under the same "green scrutiny" as many others, as Ms Ruby Wong indicates, "having gone on for centuries without any indication of environmental consciousness". Dr Wong holds a B.S. Degree in Biological Chemistry from the University of Detroit Mercy and has almost a decade of experience directing and working in research labs in materials science, environmental and regulatory compliance, quality control, quantitative analysis, and green chemistry.

She describes the industry from "the ground up" – planting, nurturing and harvesting, and includes trends in forest coverage on Earth, plant hybridization, degradation and decay, and reforestation, all with a view to environmental impact, carbon footprint, and energy impact. Sawing, drying, planning and subsequent processing are also discussed. The entire chapter is a very refreshing and enlightening view.

1.2.7 Determining wildland fire markers in residential structures using thermal desorption gas chromotography mass spectrometry

One artifact of our on-going climate changes appears to be the increased frequency and intensity of wildland fires with considerable damage and cost to residential and commercial property. Presently there are no specific industry standards or guidelines for analyzing the resulting combustion residues. In this chapter, Professor Dale LaCaptain, K. Martin, and A. Delia summarize some of these events, and their results with directed focus upon method development for analysis.

1.3 Summary

We do recommend anyone who is interested in what we have presented here to please examine the previous volumes in this series as well, to see who has authored chapters in them, and what subjects have been covered. As this series continues to expand, we welcome discussions about ideas for future volumes with an eye to welcoming your contribution. If you are interested, please contact us.

Attila E. Pavlath

2 Biodegradable polymers: Why, what, how?

Abstract: Polymer production and use is an enormous industrial concern, with a few commodity polymers constituting a large portion of the market. The degradation of these traditional polymers is extraordinarily slow, and thus a large effort has been made to produce biodegradable polymers on an industrial scale. A survey of the advantages and challenges of such novel materials are presented.

Keywords: plastics, polymers, biodegradable, renewable sourcing

2.1 Introduction

In our everyday life we encounter numerous polymers in various applications. The six most common plastics: polyethylene, polypropylene, polystyrene, polyvinyl chloride and polyethylene terephthalate are each completely non-biodegradable. Manufacturing rates of these materials are approximately 10–15 billion pounds/year [1]. Generally, we do not think about their chemical composition. To the layman their use is important regardless of how they are prepared. We are concerned about their properties, e. g. strength, life span and possible toxicity. Naturally, as chemists we want to know about how they are formed not just for scientific curiosity, but to see the ease or difficulty of obtaining the material and whether the economics justify it. However, in the past we frequently overlooked what happens with such material after it served its use. This is especially important for polymeric materials what we use in our everyday life.

Polymeric materials regardless of the ways they are obtained and/or prepared will decompose after they served their purpose. Obviously any organic polymers can be burned instantaneously, but in the absence of thermal degradation the physical and microbiological environments attacks the polymer chain and start a process to cut the chain into smaller pieces regardless of its structure. The only difference is how long this process will take in the given environment. The length of time depends on its structure. It might take only days, or many years. While such physical factors as sunshine and whether conditions can start degradation, it generally takes years. A polymer is considered biodegradable if it decomposes in a microbiological environment. The length of decomposition is dependent on the structure. There is no exact numerical value, but for practical purposes if the degradation takes many years under such conditions, the polymer is not considered biodegradable.

This article has previously been published in the journal Physical Sciences Reviews. Please cite as: Pavlath, A. E Biodegradablepolymers: Why, what, how? *Physical Sciences Reviews* [Online] 2020, 5. DOI: 10.1515/psr-2019-0020.

https://doi.org/10.1515/9783110597783-002

2.2 How degradation can occur

Generally, polymers can be grouped into two classes. One is where the main polymeric chain itself contains only carbon atoms. Other than thermal degradation the carbon–carbon bond is quite resistant to fast, everyday reactions which occur in the atmosphere or in the ground. While degradation still might occur in years, such as through ultraviolet irradiation, this places polymers with carbon–carbon chains in the non-biodegradable category. This can actually be an advantage for these types of polymers if their use is in the outside environment where they are exposed to sunlight and various weather phenomena. Polymers which have tertiary carbon atoms, e. g. polypropylene, can be affected by ultraviolet light, creating free radicals which will interact with the oxygen in the air, starting a slow decomposition. Polymers containing aromatic rings, e. g. polystyrene, are also susceptible to UV radiation. However, there is a problem. Some of these polymers are prepared with the inclusion of antioxidants to prevent degradation while in use, and these will slow down decomposition when discarded. The degradation will still take years.

On the other hand, when the polymer chain includes other atoms such as oxygen (polyesters) or nitrogen (polyamides) they are subject to possible hydrolysis which might be aided by microorganisms. The rate of the breakage of the polymer chain depends on the environment surrounding the polymer: in air or buried under ground.

One possibility of increasing biodegradability of a pure, non-biodegradable polymer is to add a component which can be consumed by microorganisms and in the process providing chemicals needed for a possible enzyme to fragment the main polymer chains. Another possibility is to incorporate auto–oxidants in order to create peroxide radicals when mixed with soil. However, again there is a problem because degradation might occur while in use. Obviously a balance has to be found between the time for a polymer's use and the start of the decomposition.

2.3 History

For their practical use biodegradable materials are most important in the medical sciences. They can be used as implants to replace bones or other body parts. In surgery they may be used for sutures. The advantage is that once the sutures have served their purpose they do not have to be removed, they will disappear by degradation. There are cases when this use is permanent, such as optical lenses, where they should be biocompatible, but not biodegradable. Obviously anything used for either of these purposes must have certain critical properties to be used in a living body, human or animal. These properties include:

1. The most important, that they should be compatible with the living organism during its use, not causing any inflammation or discomfort,
2. During their degradation they cannot release any toxic by-products,
3. They should have the necessary physical properties related to their intended use, such as strength, permeability.
4. Their rate of degradation should be in harmony with the required range of time of their application.

Biodegradable materials are used for encapsulation of medication not only to make it palatable, but also if controlled release of the encapsulated material is required. They may be used in agriculture for slow release of active components in the ground. Finally, edible films are used for numerous applications throughout the food industry; and the term "edible" indicates not only the required characteristics, but also FDA approval for consumption [2]. The simplest such product is the casing of sausages.

While biodegradable materials play an important role in medical and food applications, their increasingly major importance is on decreasing waste. In everyday life we use many polymers for various applications. Depending on the application their life times have various ranges. After there is no further need for their use, they may be disposed in various ways. Depending on the polymer they may be recycled if they can be melted down. However, this requires organized collection of the waste preferably sorted. Unfortunately, this requires everyone's cooperation, which is not done easily. Human nature, perhaps we could call it laziness, is difficult to change. This is evident in the large amount of material visible on streets, picnic places, rivers, oceans, etc. While chemistry is capable of solving various problems in our life, we are not yet at the point of changing human nature. This is where the increasing need of biodegradable polymers come in.

2.4 How to form biodegradable polymers

The major advantage of most non-biodegradable polymers is that they can be melted without decomposition and molded into a required shape. This is the major problem of most biodegradable polymers. Very few can be melted without decomposition, which will destroy their advantageous properties. Some of them may be dissolved in water and cast in some form, but the resulting material may be subject to dissolution during use. While in some cases this might be acceptable, such as encapsulation, in most applications of polymers structural stability and resistance to the environment during their utilization is required. Therefore, we have two possible approaches:
- create the required shape from solution and fortify it afterwards
- depolymerize a biodegradable polymer so it can be shaped, and then restore the long molecular chain

The first method is especially applicable for the creation of biodegradable films used for packaging. Various natural carbohydrate type of materials can provide films from aqueous solutions. Using acidic components, such as pectin and alginate, transparent films can be cast. These films can be stabilized against dissolution in water by treating them with multivalent ions forming ionomers. Such films become insoluble in water and provide films with strength similar to polyethylene.

The second method sounds possible if the depolymerization is reversible. Repolymerization such as hydrolyzed ester-type of bonding requires special conditions for recombination after forming the desired shape. There is only one type of bonding which is suitable for this process: -S-S- bonding. It can be opened by reducing agents forming -SH + HS- which can be easily converted to the original -S-S- structure and creating higher molecular weight, stable polymers. Interestingly, this was a standard process well before the need for biodegradable polymers arose.

The human hair is a keratin-based protein where the protein chains are cross-linked by -S-S- bonds, since it contains the amino acid cistein. In hair salons the hair is treated under medium temperature with special reducing agents not to cause any irreversible destruction in color, softness and other desirable properties. This softens the hair by decreasing the molecular weight of the keratin and it can be shaped into the desired format. The treatment can be used for curling or straightening the curls, it works for both. After the treatment, when conditioners are used while the hair slowly cools off, the -S-S- bond will recombine retaining the new form. Before this method was commercially applied in hair salons, the curling iron was used, and still can be. Based on this principle any polymer containing cistein can be shaped. The result is dependent on the amount of cysteine in the polymer.

2.5 Type of polymers: preparation, utility, degradation, problems polysaccharides

There has been a long history of attempts to utilize starch as well as other carbohydrates to create biodegradable plastics, in a variety of formulations. The major technical problem overall is the need to optimize two characteristics that can be at odds: stability of the polymer and its moldability. Two important factors in such formulations are water content and molecular weight. Whatever polymer is produced must, after its intended use, be biodegradable; however, the Polymer must be Inert and robust over its intended lifetime and for its intended application. Such stability routinely involves both low water content as well as resistance to water. Of note however, most agricultural polymers require water while undergoing the molding process, as a plasticizer. Better stability and physical strength are gained with high molecular weight materials, yet this decreases overall moldability.

Starch is perhaps obviously the most thoroughly examined biodegradable natural polymer. Starch is potentially thermoplastic, tends to be a relatively low-cost polymer, and has agricultural surplus as its major feedstock or source. It is processed with water however, and this water content ultimately in the product does affect the final stability of the polymer.

Starch is an atypical thermoplastic polymer, because high temperature and pressure, as well as moisture can be used in its processing. Inorganic salts, including NaCl and $CaC1_2$ can alter and affect the thermoplasticity of starch. The use of these as additives do reduce any product's hydrophilicity, yet must utilize a higher processing temperature [3]. Relatively low [4] and high [5, 6] concentrations of starch have in the past been incorporated as fillers in non-biodegradable polymers.

Cellulose is a second widely available carbohydrate. While it is biodegradable, unfortunately it cannot be processed through thermal means, since it decomposes prior to melting. Improvements in thermoplasticity can be made through chemical modifications, but this is accompanied by decreases in biodegradability. Biodegradation generally does not occur if what is called the degree of substitution (abbreviated DS) is significantly above 1.0. DS is defined as a ratio of number of substituents per glucose unit. Values range from a low of zero to a high of 3.0. It has been determined with cellulose ethers [7–9] as well as with cellulose acetate [10].

It is theorized that the substitutions create both steric hindrance and decreased hydrophilicity. Slower biodegradation is the end result.

Two types of polymeric amino carbohydrates are chitin and chitosan, major components of crustacean exoskeletons – shells – and insect wings. The microcrustacean krill, and their harvesting, become a source of chitin, although waste materials high in chitin can also be gathered and used [11]. Both types of material are promising polymers for use in films. Each is resistant to the transmission of oxygen through the material, and has, outstanding mechanical properties. Studies of their reactions [12] and biodegradability [13] are well established.

Another polysaccharide material, chitosan, is structurally very like cellulose. The compound exists as a polymeric form of 2-amino-2-deoxy-n-glucose including the 1,4-glucosidic linkage. Structurally, chitosan is affected by the following factors: crosslinking agents plus degree of crosslinking, interactions with components such as salts with which it comes into contact, various acids with which it is neutralized, and the degree of water activity of the matrix. An example of structural sensitivity may be deduced via the end result of the use of nonporous chitosan-based membranes during what is known as the pervaporative separation of water from various mixtures of ethyl alcohol and water. Pervaporation is a process by which a chitosan matrix effects passage of water preferentially through the integration of selective adsorption from an ethanol-water solution, as well as selective solubility within the matrix, plus differential diffusion throughout the matrix, and also via vapor phase selective desorption, this phase subsequently being condensed. Cobalt, manganese, and zinc sulfates, when in contact with the pervaporative membrane, change the selectivity for water. This is because "holes" in the structure contract due to complex formation occurring involving the salt cations.

Several different types of long-chain fatty acids can be combined with chitosan solutions in 1% aqueous solutions of small molecular weight carboxylic acids, such as formic and acetic acid. The formation of salts takes place after evaporation, the volatile small molecular weight formic or acetic acid being replaced with the less volatile higher molecular weight fatty acids. The water barrier properties of the resulting films are dependent upon the acids used. It is lauric acid that has exhibited optimal water resistance [14].

While such film alone has moderate, sometimes good, strength bonding such films to paper can be increase this. Since chitosan has found use as a paper additive, resulting in increased overall strength, films can be bonded to paper, either a single sheet or a sandwich between sheets through wetting. Non-biodegradable non recyclable polyethylene-coated wrappers can potentially be replaced by coated papers such as this. Rates for the transmission of water are not as good as already existing commercial products. Starting substances with similar properties may be possible through the successful amination of starch and other carbohydrates from agricultural feedstocks.

2.6 Pectine and alginic acid

These are carbohydrates with acidic carboxyl groups which readily form salts. They can be cast to form transparent films, however after the evaporation of the water they can be subject to aqueous re-dissolution or loss of their favorable characteristics through exposure to high humidity. Cross-linking the film surface will alleviate such degradations of the material. Various reactions can be achieved through covalent bonding, for example, treatment with formaldehyde. This may however result in novel chemical architectures, which in turn can become a problem if the films will ultimately be used for food purposes, and may not be approved by FDA. Interactions with multivalent ions, and the formation of ionomers is currently the most productive reaction method. Although most synthetic films possess higher tensile strengths than readily encountered edible films, ionomers do not.

The general method is to cast the water soluble alginate or pectin to form a film and then immerse the dry film into a multivalent ion solution. At that time two competing reactions will take place. One is that the water will penetrate into the dry film to dissolve it. The other reaction that the multivalent ion will interact with the carboxylic group forming an insoluble salt, thus preventing the dissolution of the film. The reaction conditions and the characteristics of the multiple ion will decide the outcome. It was found that Ca^{++} and Cu^{++} provided the most insoluble films. At the same time, Ca^{++}, Cu^{++} and Zn^{++} treatment not only tripled the tensile strength of the untreated films, but they surpassed by 50% that of commercial cellulose acetate and by 100% the polyethylene film [15, 16].

2.7 Protein

Proteins tend to be somewhat more expensive than carbohydrates, but are the more easily biodegraded of the two classes of material. Some medical settings already see collagen films utilized [17]. This animal protein product, collagen, can undergo partial hydrolysis. The lower molecular weight product polymers, called gelatins, are used widely in both the food and pharmaceutical industries. Despite cross-linking, gelatin remains biodegradable [18].

Casein derived from milk is a further protein that is readily available. Casin-based films do transmit moisture, yet have excellent mechanical properties in many applications [19]. When casein, carbohydrates, and fats are combined, reduced water vapor transmission is observed in the resulting films [20].

Various proteins containing cysteine can provide transparent strong films, as previously addressed. Keratin, which is abundant in human and other mammalian hairs, contains disulfide bonds. Such bonds are readily subject to redox reactions. The disulfide bonds can, through cross-linkings, result in higher molecular weights, and can be opened or re-opened via reducing agents as simple as sodium bisulfite. The peptide bonds are unaffected by such reactions. The fibers do become pliable enough to result in films under optimal conditions, when such crosslinking is temporarily opened. Wool fiber treated with sodium bisulfite and pressed at 130–140 °C provided strong transparent hydrophobic films. Such films had limited transparency under 130 °C [21, 22].

Using chicken feather, which does includes a keratin component, although a less cross-linked form than wool, homogenous films were produced, of varying opacity, under elevated temperature conditions (75–110 °C in 3 min), with no improvement above this. A further factor affecting the formation of films, pressure, was found to be a major characteristic in the clarity of the films produced, with a maximum of 46.7 MPa being optimal. The best results, the most transparent films, resulted when the feather was ground mechanically, using some mixer [23].

Keratins, regardless of their origin have high amount of crosslinking through cistein. However, it is not the amount of crosslinking, but rather their location is the deciding factor in film formation. Wheat gluten contains only 1.3–1.8% of cistein. Using various inorganic reducing agents as for wool and chicken feather, transparent strong films were obtained at 110 °C under pressure. At lower temperature longer times were needed and at higher temperature discoloration started to occur [24].

2.8 Polyesters

Polyesters are entirely synthetic polymers universally produced from dicarboxylic acids and diols [25]. Most of them require 200 years or more to biodegrade. Within this broad class of synthetic materials, the aliphatic polyesters are almost the only biodegradable compounds of high molecular weight [26].

Poly(butylene succinate) and poly(ethylene succinate) are examples.

They are generally prepared by the polymerization of hydroxyacids, HO-R-COOH or from cyclic lactones. These polymers are more easily biodegradable because of their hydrolyzable ester bond, but their properties are not as good [27]. Examples include:

Polyglycolide (PGA): the simplest linear aliphatic polyester produced by the polymerization of glycolide, which is a cyclic lactone of the glycolic acid. Melting point is around 220 °C [28]. It has good mechanical properties, but its degradation results in acids which limits it biomedical application. It requires copolymerization with lactide or caprolacton for such use [29].

Polylactide (PLA): It is prepared from L-lactic acid and lactide 14.

It has a melting point of 170 °C. However all chemical properties, even biodegradability, can change depending on the optical form of the lactic acid (D, L or DL) used for polymerization [30].

2.9 Can we save the environment?

Polymers serve an important role in various applications to make our everyday life easier and more enjoyable. As was pointed out at the beginning, the problems start when they are no longer needed for their given purpose. Some of them can be recycled or if not, they can be buried in garbage dumps. The latter is certainly not perfect, but at least would not represent an eyesore when they are carelessly thrown away. Recycling and collection require continuous careful attention by everyone. Changing human nature and behavior probably would take more time than the biodegradation of the polymers. We have this type of problem not just with polymer waste, but with other factors in our life. We seek quick solutions and do not think about future problems related to the solution. For example, suggestions are made for solving energy problems by conserving energy in daily life. Organic agriculture is thought to be solution for food safe from chemicals. Other politically correct solutions are recommended for other problems. The overall problem is that these are only band-aids for the immediate problems. In each case we try to develop ways to address the present problem. The new solutions generally sound very enticing but ignore the economic problems they in turn create for which we will have to find new solutions, thus becoming a never ending cycle. Very few scientists speak up openly that Earth has limited resources of everything, while under the given conditions the users of these resources are increasing in an uncontrolled very rapid manner. Unless we develop the opportunity for mass emigration to other parts of the universe, we must face this issue and debate openly this problem and develop solutions. Even though that would be worth multiple Nobel Peace Prizes, very few politicians and other influential persons make this their major goal. Yet everything else is just sweeping the problems under the rug. We must use common sense, but unfortunately in the discussion of such problems common sense is not common.

References

[1] Johnson R. An Overview of Degradable Plastics. J Plastic Film Sheet. 1988;4:155–70.
[2] Pavlath AE, Orts W. Edible films: why, what and how! In: Embuscado ME, Huber KC, editors. Edible Films and coatings for food applications. New York, NY: Springer, 2009:40 pages.
[3] Huang AS, Liu XX. In: Ching C, Kaplan D, Thomas E, editors. Biodegradable Polymers and Packaging. Lancaster, PA: Technomic, 1993:53–61.
[4] Narayan R. In: Chum HL, editor. Assessment of biobased materials Report SERI/TR-234-3610. Colorado: Solar Energy Research Institute, 1989:7.1–7.25.
[5] Otey FH, Westhoff I. EC products research and development. American Chemical Society, 1984:284–7. Washington, DC.
[6] Otey FH, Doane WM. Proc. of symposium on degradable plastics. Washington, D.C.: The Society of Plastics Industry, Inc., 1987:39–40.
[7] Kasulke U, Dautzenberg H, Polter E, Phillipp B. Enzymatic degradation of some cellulose derivative. Cellulose Chem Technol. 1983;17:423–32.
[8] Perlin AS, Bhattacharjee SS. Enzymatic Degradation of Carboxymethylcellulose and Other Cellulose Derivatives. J Polym Sci. 1971;36:509–21.
[9] Wirick MG. A Study of the Enzymic Degradation of CMC and Other Cellulose Ethers. J Polym Sci Part A-1. 1968;6:1965–74.
[10] Stutzenberger F, Kahler G. Cellulase Biosynthesis During Degradation of Cellulose Derivatives by Thermomonospora Curvata. J Appl Bact. 1986;61:225–33.
[11] Naczk M, Synowiecki J, Sikorski ZE. The Gross Chemical Composition of Antarctic Krill Shell Waste. Food Chem. 1981;7:175.
[12] Pavlath AE, Wong DS, Robertson GH. In: Salamone J, editor. Polymer encyclopedia, vol. 2. Boca Raton: CRC Press, 1995:1230–4.
[13] Hirano S, Tsuchida H, Nagao N. N–acetylation in Chitosan and the Rate of its Enzymic Hydrolysis. Biomaterials. 1989;10:574–6.
[14] Wong DW, Gastineau FA, Gregorski KSJ, Tillin SJ, Pavlath AE. Chitosan-Lipid Films: Microstructure and Surface Energy. J Agric Food Chem. 1992;40:540–4.
[15] Pavlath AE, Voisin AR, Robertson GH. Pectin-based biodegradable water insoluble films. In: Cheng HN, editor. Application of polymers in foods. Washington D.C.: American Chemical Society, 1999:107–15.
[16] Pavlath AE, Gossett C, Camirand WM, Robertson GH. Ionomeric films of alginic acid. J Food Science. 1999;64:61–3.
[17] Bogdansky S. In: Chasin M, Langer R, editors. Biodegradable polymers as drug delivery systems Drugs and the pharmaceutical sciences. New York: Marcel Dekker, 1990:231–59.
[18] Huang SJ, et al. Design, synthesis and degradation of polymers susceptible to hydrolysis by proteolytic enzymes. In: Sharpley JM, Kaplan AM, editors. Proc. of the third international biodegradable symposium. London: Applied Science, 1976:731–41.
[19] Grohmann K, Himmel M. Assessment of biobased materials Report. In: Chum HL, editor. Assessment of biobased materials Report SERI/ TR- 234-3610. Colorado: Solar Energy Research Institute, 1989:11.1–11.10.

[20] Krochta JM, Hudson JS, Camirand WM, Pavlath AE, Proc. American Society of Agricultural Engineers, Chicago, 1988, December, No. 88-6523, 1–15.
[21] Pavlath AE, Lee KS, Ward WM. Symposium on membranes MACRO-20, Vol. I. Mexico City, Mexico: North American Chemical Congress, 1975.
[22] Pavlath AE, Houssard C, Camirand W, Robertson GH. Clarity of films from wool keratin. Textile Res J. 1999;69:539–41.
[23] Pavlath AE, Levy S, Robertson GH. Films from feather keratin. In: Cherry JP, Pavlath AE, editors. Proc. of 31st meeting of UJNR protein panel. Monterey, CA, 2002.
[24] Pallos FM, Robertson GH, Pavlath AE, Orts WJ. Thermoformed wheat gluten biopolymers, J. Food and Agricultural Chemistry. 2006;54:349–52.
[25] Okada M. Chemical synthesis of biodegradable polymers. Progr Polym Sci. 2002;27:87–133.
[26] Chandra R, Rustgi R. Biodegradable polymers. Progr Polym Sci. 1998;23:1273–335.
[27] Lofgren A, Albertsson AC, Dubois P, Herome R. Recent advances in ring opening polymerization of lactones and related compounds. J Macromol Sci Rev Macromol Chem Phys. 1995;35:379–418.
[28] Nair LS, Laurencin CT. Biodegradable polymers as biomaterials. Progr Polym Sci. 2007;32:762–98.
[29] Middleton JC, Tipton AJ. Synthetic biodegradable polymers as medical devices. Med Plastics Biomater Mag. 1998;3:30.
[30] Auras R, Harte B, Selke S. An overview of polylactides as packaging materials. Macromol Biosci. 2004;4:835–64.

Rose Ryntz and Susan Kozora

3 Sustainable materials in automotive

Abstract: The use of plastic parts in the production of automobiles has a long and established history. In the more recent past, manufacturers have become interested in bio-based materials for a variety of parts, in part because they represent a more sustainable material stream, and in part because they may be more economically feasible. This chapter discusses the evolution from fossil fuel-based materials to bio-based materials in their established and potential uses for automotive parts.

Keywords: automotive, plastics, recycling, bio-based materials

3.1 Introduction to plastics in automotive

Plastics manufacturing is the third largest sector of US manufacturing and accounts for about 3 % of the global economy, with an estimated value of $294 billion. Over the next 10 years, approximately 3.5 million manufacturing jobs will be created but with the retirement of baby boomers and the lack of attractiveness of the manufacturing industry, a major skill gap will result in the inability to attract over half of those numbers by 2025.

Automotive assembly plants struggle to keep up with volume, with over 17 million vehicles produced in plants running at maximum capacity. The split of sedans to sport utility vehicles (SUV) and cross-over utility vehicles (CUV) continues to increase in favor of the SUVs and CUVs. Economic and demographic forces continue to converge in conjunction with government demands for lighter, competitively priced vehicles.

In 2016, the average vehicle utilized approximately 330 pounds of plastic, approximately 8.3 % by weight [1]. Vehicles today are comprised of approximately 50 % plastics by volume and plastics account for approximately 10 % by weight [2]. By 2020, it is estimated that approximately 110 billion pounds of plastic will be used in automotive, or roughly 18 % by volume.

Plastics are being used in incredibly innovative ways to help make cars safer and more fuel efficient. They provide the solutions that make vehicles more lightweight, help increase fuel efficiency and reduce carbon emissions, help provide unparalleled safety benefits like airbag systems and safety curtains, and help to enable beautiful, crafted components and sleek designs. What's more, as car makers work to meet sustainability goals and use recovered materials, a growing number of car parts are made using recycled plastics.

This article has previously been published in the journal Physical Sciences Reviews. Please cite as: Ryntz, R., Kozora, S. Sustainable materials in automotive *Physical Sciences Reviews* [Online] 2020, 5. DOI: 10.1515/psr-2019-0021.

https://doi.org/10.1515/9783110597783-003

3.2 Key material attributes/specifications for use in components

The low density of plastics affords components that are lighter weight and makes for more fuel- efficient vehicles. It is estimated that every 10 % reduction in vehicle weight results in a 5 % to 7 % reduction in fuel usage [3]. Current economic and environmental concerns make the creation of more fuel- efficient cars a top priority in the automotive industry. Some other advantages of high-performance plastics include:
- minimal corrosion, allowing for longer vehicle life
- substantial design freedom, allowing advanced creativity and innovation
- flexibility in integrating components
- safety, comfort and economy
- recyclability

While there are generally 13 high-performance plastics that make up the majority of plastics utilized in automotive components, just three types of plastic make up approximately 66 % of the total high-performance plastics used in a car: polypropylene (32 %), polyurethane (17 %) and PVC (16 %) (see Table 3.1) [4].

Poly(propylene) is a thermoplastic polymer used in a wide variety of applications. A saturated addition polymer made from the monomer propylene, it is rugged and

Table 3.1: Plastics Used in Automotive Applications.

Plastic Type	Abbreviation	Automotive Applications
Poly(propylene)	PP	bumpers, instrument panels, door panels, carpet fibers
Poly(urethane)	PUR	flexible foam seating, foam insulation panels, elastomeric coatings, instrument panel foam, door foam buns
Poly(vinyl chloride)	PVC	instruments panel skins, sheathing of electrical cables, door skins and bilaminates
Acrylonitrile Butadiene Styrene	ABS	body parts, instrument panels, door panels, wheel covers
Poly(styrene)	PS	housings, buttons, display bases
Poly(ethylene)	PE	fuel tanks, films, electrical insulation
Poly(oxymethylene), Acetal	POM	interior and exterior structural parts, fuel systems, small gears
Poly(carbonate)	PC	bumpers (with co-polymer), headlamp lenses
Poly(methyl methacrylate)	PMMA	windows, displays, screens
Poly(butylene terephthalate)	PBT	door handles, bumpers (with co-polymer)
Poly(ethylene terephthalate)	PET	wiper arm and gear housings, headlamp retainer, engine cover, connector housings
Acrylonitrile Styrene Acrylate	ASA	housings, profiles, interior parts

unusually resistant to many chemical solvents, bases and acids. It can be compounded with other ingredients to create multi-functional components. For example, when compounded with long glass fibers at weight percentages of 20–30 % the resultant component utilizing such a compound can be utilized for structural instrument panel retainers. If compounded with elastomers and fillers, the resultant material, referred to as a thermoplastic olefin (TPO), can afford high-impact resistant bumpers.

Poly(urethane) is an *elastomeric material* of exceptional physical properties including toughness, flexibility, and resistance to abrasion and temperature. Poly(urethane) has a broad hardness range, from very soft and flexible to very hard. Other polyurethane characteristics include extremely high flex-life, high load-bearing capacity and outstanding resistance to sunlight (if aliphatic in nature) ozone, radiation, oil, gasoline and most solvents. Poly(urethane) foams are utilized in instrument panels and door rollover uppers to provide a flexible haptic. These foams are generally foamed in place between the substrate and the coverstock via a reaction injection process of the two-component isocyanate and polyol. The foams on door and center console armrests are often applied as a foam bun that is previously made and cut to shape.

Poly(vinyl chloride) (PVC) has good flexibility, is flame retardant, and has good thermal stability, a high gloss, and low (to no) lead content. PVC molding compounds can be extruded, injection molded, compression molded, calendered, and blow molded to form a huge variety of products, either rigid or flexible depending on the amount and type of plasticizers used. When used as a cover-stock skins in instrument panels or door panels the PVC is plasticized with large amounts of plasticizers, generally in the 40 to 60 weight percentage range. The plasticizers, which are synthetic (odd number of carbon atoms in the molecular chain) or natural (even number of carbon atoms in the molecular chain), if of too low molecular weight, will often migrate from the skin over time and lead to a brittle coverstock. The migration of plasticizers is what leads to the "fog" seen over time on windshields. PVC can also be constructed into a bilaminate coverstock which is comprised of a thin PVC skin laminated to a urethane or cross-linked poly(propylene) foam.

Acrylonitrile Butadiene Styrene is a copolymer made by polymerizing styrene and acrylonitrile in the presence of polybutadiene. The styrene gives the plastic a shiny, impervious surface. The butadiene, a rubbery substance, provides resilience even at low temperatures. A variety of modifications can be made to improve impact resistance, toughness, and heat resistance.

Poly(amide), for example, Nylon 6/6, is a *general-purpose nylon* that can be both molded and extruded. Nylon 6/6 has good mechanical properties and wear resistance. It is frequently used when a low cost, high mechanical strength, rigid and stable material is required. Nylon is highly water absorbent and will swell in watery environments. Nylon must be dried prior to being molded or surface defects, known as splay, will result. The resultant component, if molded with internal moisture, will also suffer in mechanical properties. If compounded with glass, a very high modulus, structural component can result.

Poly(styrene) is a naturally clear polymer which exhibits excellent chemical and electrical resistance. Special high gloss and high-impact grades are widely available. This easy to manufacture plastic has poor resistance to *UV light*, degrading over time which result in stress cracks.

Poly(ethylene) has high impact, low density, and exhibits good toughness. It can be used in a wide variety of thermoplastics processing methods and is particularly useful where moisture resistance and low cost are required.

Poly(oxy methylene), POM which is also called acetal, has excellent stiffness, rigidity, and yield strength. These properties are stable in low temperatures. POM also is highly chemical and fuel resistant. It can be easily blended with colorants to resultant in scratch resistant mold in color components.

Poly(carbonate) is an amorphous polymer which offers a unique combination of stiffness, hardness and toughness. It exhibits excellent weathering, creep, impact, optical, electrical and thermal properties. Because of its extraordinary impact strength, it is a material that is often used in bumpers and instrument panels when compounded with other elastomers. A major issue with poly(carbonate) is its lack of resistance to chemicals, particularly bases, which when exposed to elevated temperatures and humidity, will hydrolytically stress crack and lead to a degradation of material properties, in particular impact resistance.

Poly(methyl methacrylate), PMMA, is an optically clear material which is often used as a lightweight or shatter-resistant alternative to glass. It's cheaper than poly(carbonate) but is also more prone to scratching and shattering.

Poly(butylene terephthalate) is a thermoplastic which is used as an insulator in the electrical and electronics industries. It is highly chemical and heat resistant. *Flame-retardant* grades are available.

Poly(ethylene terephthalate) is mostly used to create synthetic fibers and plastic bottles. It is referred to under the name "polyester."

Acrylonitrile styrene acrylate, ASA has great toughness and rigidity, good chemical resistance and thermal stability, outstanding resistance to weather, aging and yellowing, and high gloss. If burned it will create cause a toxic smoke.

The automotive interior segment continues to consolidate and be driven to new and complimentary technologies to address future automotive trends. The connected car is expected to be a $50 billion market in 2020, with 33.3 million autonomous vehicles produced in 2040, with over 85 % of them being powered by an electric drivetrain [5]. The interaction between the car and the driver is becoming paradoxically more complex, with less time to drive and more time to monitor. The human-machine interface (HMI) will become simpler, easier to understand from the 150 devices used today to regulate safety, comfort, and convenience. The lifestyle demands of sight, sound and touch, will drive advances in personalization (3D printing), illuminated lighting, haptically pleasing surfaces, and acoustic performance.

Understanding the variance in demographics globally will be the key in technology introduction and shortening the time of new product development. North America will continue to demand greater utility and space while Asia will drive the need for better air quality, lower volatile organic content (VOC) and lower odor in the vehicle interior, rear seat utility and lighter colors will be the norm. Europe, on the other hand, will mandate personalization, well-embedded components that exhibit craftsmanship and fit and finish. The use of premium surfaces, two-tone colors, and augmented premium surfaces with leather or synthetic leather haptics will be used.

All the while environmental demands, clean and green, carbon credits, disposal vs re-use and recycling, will continue to be promulgated. With the increased use of multi-materials, increased electronic content, and the drive to lower odor components, sustainability and the use of more natural vs. petroleum-based solutions may be mandated.

3.3 Sustainable material requirements

3.3.1 Why sustainable materials?

It's the key question – why are we looking at sustainable materials? Is the current set of materials typically used in automotive plastic components considered sustainable?

Merriam-Webster defines sustainable as:

adjective: of, relating to, or being a method of harvesting or using a resource so that the resource is not depleted or permanently damaged

Sustainability encompasses the total environmental, economical and social impact of materials.

When considering the plastics used to make an automobile, determining the source of the building blocks, or monomers, helps to define its level of sustainability. The level of complexity and variety in the types of plastics and polymers used in a vehicle makes it a challenge to provide sustainable solutions and environmentally friendly alternatives as well as the capability to separate all of these different materials for reuse and recycling. It also illustrates the need for design for environment and recycling as early as possible in the vehicle conception phases as the whole vehicle is brought together. Because of the many polymers/plastics used today, sustainable solutions are being sought with a wide variety of technologies.

The monomers used to create most of the polymers listed come from either natural gas or petroleum sources via cracking and/or distillation. This being a finite source does not fit the definition of sustainable materials since the source, once harvested, is depleted.

With the automotive industry contributing 3.04% of the US gross domestic product (GDP) in 2017 [6] and a growing market in developing countries, the need for sustainable materials is accelerated by the size of this industry and its growth. An industry as large as the automotive has a responsibility to the environment and the use of global natural resources.

Solutions to increase the use of sustainable materials point towards bio-based and recycled options. Bio-based plastics/reinforcements use a renewable, “new carbon” resource for its formulation/structure. Bio-based materials may also reduce a product’s overall carbon footprint. Since many types of plastics can be recycled over and over again, the initial stream of material is not depleted and can be used repeatedly. While not as renewable as being derived from plants, recycled materials reduce use of resources, fossil fuels, landfill, carbon footprint and costs (in most cases). They provide a more stable business case as their pricing is comprised mainly of transport of the feed stream, supply of feed stream and reprocessing/compounding. The effect from potentially volatile fossil fuel costs or climate effects on crop yield and quality is minimized or non-existent with recycled products.

3.3.2 How do we define bio-based materials?

3.3.2.1 How does a material qualify as biobased?

“Green washing” is a term used when referring to misuse of facts and information given to the general population regarding the environmental attributes of a product or material. If there is little understanding of what truly defines a sustainable material, then it becomes too easy to confuse and misdirect, especially with respect to marketing a product.

For example, considering the term plant-based polymers or bio-polymers, one would think of a sustainable product. However, one can also say that natural gas or crude oil as fossil fuels, come from plants and living creatures – organic matter-from millions of years ago. So, technically, even plastics made from such beginnings can be considered plant or bio-based! Therefore, we need to focus on renewably sourced materials.

As previously defined, sustainable materials must come from sources that are not depleted or damaged, in other words, renewable sources. For those sourced from natural gas or petroleum, those resources are consumed without replacement. Sustainable materials can then be considered as those from renewable origins such plants grown in crops or that have a relatively reasonable life cycle to keep up with demand without concern for depletion and can be renewed or replaced by new growth or reuse of the material after intended end of life usage as in recycling.

3.3.2.2 Verification and certification methods

In order to determine from which sources monomers are derived – since, ideally, polymers made from sustainable versus finite elements should have the same structure and properties – methods must be used to distinguish the origins of the materials. Thus there are several standards globally to determine the level of "new carbon" to define bio-based content of a material.

For organic (carbon based) polymers this is done by looking at the carbon element in the material. All of the plastics used in automotive applications are carbon-based with the exception of silicone-based adhesives and sealants, but those materials are relatively insignificant in quantity and are not considered here.

Radioactive carbon dating is used to assess the bio-based content of organic materials to differentiate "old" (fossil fuels) vs "new" (renewable bio sources) carbon. In the US, the method most commonly used is American Standard Test Method -ASTM D6866 [7] which is the standard analytical method for measuring the bio-based content of a product.

Per the ASTM D6866:

> This standard is a test method that teaches how to experimentally measure bio-based carbon content of solids, liquids, and gaseous samples using radiocarbon analysis. These test methods do not address environmental impact, product performance and functionality, determination of geographical origin, or assignment of required amounts of bio-based carbon necessary for compliance with federal laws. [7]

This method is us by the United States Department of Agriculture (USDA) for their Bio Preferred Program [8] which prioritizes use of bio-based products for US government purchased products. This program also provides bio-based content certification information, definitions and certified suppliers of bio-based materials to the public.

Globally, other methods applying the same type of carbon dating are used:

- European standard [9] – EN 16785–1, *Bio-based products – Bio-based content – Part 1: Determination of the bio-based content using the radiocarbon analysis and elemental analysis* – EN 16785–2, *Bio-based products – Bio-based content – Part 2: Determination of the bio-based content using the material balance method.*
- International Standards Organization [10] (ISO) – ISO 16620–2. *This part of ISO 16620 specifies a calculation method for the determination of the bio-based carbon content in monomers, polymers, and plastic materials and products, based on the 14C content measurement.*
  This part of ISO 16620 is applicable to plastic products and plastic materials, polymer resins, and monomers.

There are also several international organizations which provide certification of materials as bio-based for regulatory, branding and marketing purposes. Below are some examples:

- International Sustainability and Carbon Certification- www.iscc-system.org
- DIN CERTCO (Germany) – www.dincertco.de
- TÜV Austria Belgium (previously Vinçotte)- www.okbiobased.be
- EcoLogo/UL (Canada) – www.ecolabelindex.com/ecolabel/ecologo
- Japan BioPlastics Association's BiomassPla – www.jbpaweb.net/english/english.htm
- Japan Environment Association Eco Mark Program – www.ecomark.jp/english/
- US Department of Agriculture Bio-Preferred Program – www.biopreferred.gov
- US Environmental Protection Agency Safer Choice Label- www.epa.gov/saferchoice/learn-about-safer-choice-label
- Sustainable Biomaterials Collaborative – www.sustainablebiomaterials.org

It should be noted that bio-based content is one measure of sustainability, albeit a significant factor. Other considerations such as resource consumption (land, water, energy, etc.), chemical toxicity, food sources and genetic modification can be included in defining a material as sustainable.

3.3.2.3 Examples of materials defined as bio-based

There are several examples of sustainable, bio-based materials for replacements of the wide variety plastics listed earlier, and a few examples will be listed here. The sustainable aspect of a plastic can be improved by changes in the polymer or by changes in or the addition of bio-based fillers. In some cases both the polymer and the filler/reinforcement can be bio-based. A plastic derived from fossil fuels can contain bio-based fillers which provide a portion of the volume of the compound and improve the overall sustainable nature of the compound.

Before moving onto examples, it is extremely important to understand that if a material is bio-based, it does not mean that the material is also bio-degradable or compostable. This has been a huge misconception as a lot of the publicized development of bio-based polymers in recent years has involved compostable/degradable materials such as poly(lactic acid) (PLA) which is a compostable polymer by its structure. It should be understood that the *source* of the monomer used to create a polymer is what defines it as sustainable or not. It does not, however, define its bio-degradability or compostability. Bio-degradable or compostable materials depend on which monomers are used and how they are polymerized, regardless if they come from fossil fuel or renewable carbon sources. So, yes, a fossil fuel-based polymer can be designed to be bio-degradable and a bio-based polymer can be designed to be durable! This is significant as for automotive applications durable plastics are required to meet the expectations of 150,000 mile/10 year long term testing under stringent conditions.

Considering renewable sources for either the polymer or a filler/reinforcement or both moves the materials used for automotive applications towards more sustainability.

Examples of bio-based plastics:
- Poly(urethanes) – PU (foams and adhesives) – Polyol is one of the basic monomers in PU formulations and can be made from plant oils such as soybean or castor bean.
- Poly(amides) – PA (nylon) – Monomers from plant oils plant oils such as soybean or castor bean.
- Poly(ethylene) (PE) – Ethylene monomer from sugarcane.
- Poly(lactic acid) (PLA) /Poly(propylene) alloys – PLA portion is based on renewable resources such as corn starch (in the United States), tapioca roots, chips or starch (mostly in Asia), or sugarcane (in the rest of the world) [11].
- Poly(trimethylene terephthalate) Polyester (PTT) – made from plant sourced glucose which is used to derive the propanediol which is then reacted with terephthalic acid. Dupont's **Sorona**® contains 37% bio-based content and is an excellent and durable engineered PTT plastic [12].
- The chemistry to get from the plant or renewable source to the base monomer, a polyol, in this case is illustrated in Figure 3.1 [13]:

Examples of bio-based reinforcements:
- Natural fibers: Kenaf, hemp, jute, flax, bamboo, agave, coconut
- Cellulose (trees): fibers, powders
- Agricultural waste: rice or oat hulls, wheat straw, flax fiber

Regarding the plant-based source of the monomer, much debate has risen around the subject of use of food crops or farming land to produce the feedstock for bio-

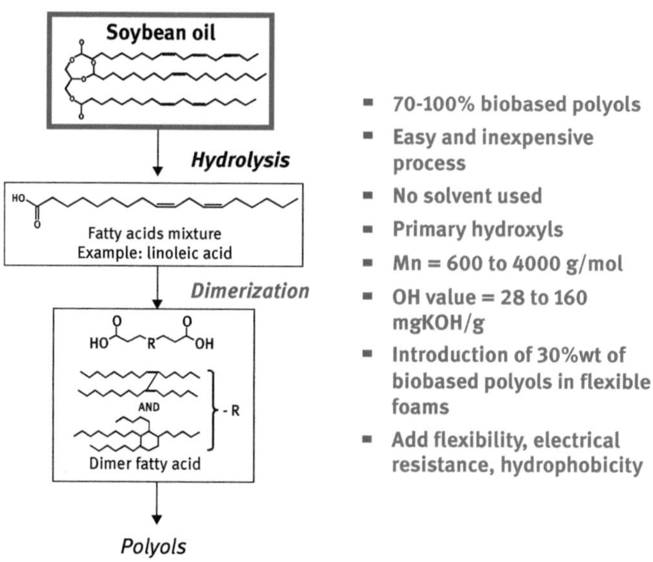

Figure 3.1: Biobased Polyl Production. Dr. Ramani Narayan, Michigan State University

based materials. Use of corn for ethanol as fuel has received criticism and some countries do not support use of food-based bio-polymers. To comply with this concern, use of agricultural waste as the feedstock is used or non-food source plants. Examples of agricultural waste include wheat straw, flax fiber, corn stalks, agave fibers, coconut fibers, and soy meal protein from soy bean oil production. Non-food sourced plants examples are fibers from kenaf, hemp, jute and bamboo fibers and cellulose reinforcements from trees. In some cases, basic monomers come from the gases evolved from composted materials as with the PHA polymer developed by NewLight Technologies [13].

3.4 Recycled materials and methods

3.4.1 Types of materials currently recycled in automotive

The automobile is one of the most recycled consumer product on the market. Not only are its components and materials recycled at a rate of 80–90 % or more of the total vehicle weight, automobiles use a significant amount of that material back into new vehicles, especially in terms of the metals used [14].

With respect to the use of recycled plastics, below illustrates use of recycled plastics materials throughout the vehicle. Although the list is extensive, there is still quite a bit of opportunity to increase the use of recycled materials in the automobile.

3.4.2 Methods used to recycle

There are several methods used to recycle plastics that can and are used in the automotive arena. First, it is best to distinguish the two major sources for reprocessed plastic compound feedstock: Post-Industrial Recyclate and Post-Consumer Recyclate.

3.4.2.1 Post-industrial recycling

Post-Industrial Recycling (PIR) is defined as reuse of scrap or excess materials used in the manufacturing process of a product. This material has not yet reached the end user and is generated at the related manufacturing facility. PIR plastics can be sourced from the virgin resin manufacturing, plastic part manufacturing, plastic part/end item assembly or packaging materials.

PIR sources of plastics tend to be more consistent and better defined due to the controlled exposure the materials have in the manufacturing process. This type of feedstock has limited exposure to the elements and "wear and tear" from use.

If the post-industrial source is from the virgin resin suppliers, this may be in the form of transition material. This refers to the resin generated when changing from one grade of material to another. Since many virgin raw material extrusion lines run continuously, the inefficiency of stopping a line to change formulations outweighs the loss in transition materials. As the lines are monitored for inputs of raw materials and quality outputs, the material generated in between formulations is sometimes known as "wide spec" materials. The term comes from the wider range of properties in the resin due to the change in formulations. These recycled material feed streams become the basis of recycled compounds providing good quality resins with stable costs.

For the most part, the wide spec materials are great base material for the recycled compounds and they can be down-cycled or up-cycled if they are blended with virgin resins or additives.

When the PIR source is from the part manufacturing or part/end item assembly, the major issue here is proper separation of the different plastics families (PP, ABS, PC/ABS, PA, etc.) within the manufacturing facility, or if a scrapped assembly (such as a fully assembled instrument panel or door for example) is dismantled. Scrap parts can be generated due to processing issues (short shots, contaminants), color change transition parts, or damage during movement/packaging or secondary processing. There may be other potential contaminants such as paint, adhesives and tapes. Manufacturing related materials such as process oils and release agents can be problematic in this type of feedstock.

Another source of PIR is in the scrap materials generated in the part molding process. Injection molding can produce sprues and runners, which essentially are the channels in the injection process that carry the molten plastic to the tool cavity which forms the part. In some cases, with multi-cavity tools with small parts, the weight of the runner can be greater than the weight of the molded parts!

3.4.2.2 Post-consumer recycling

Post-Consumer Recycling (PCR) refers to plastics recovered after the consumer has used the product. Examples would be recovered automotive parts, curbside recycled plastics, carpet, water/soda pop bottles and recycled electronics.

One use or short-term use plastics, such as plastic wear – cups, utensils, plates – have gained notoriety lately especially with respect to plastic bags and plastic straws. Some states are banning use of either as an attempt to combat the trash issue and minimize environmental impact.

When plastic ends up in the ocean, ground, and landfills, it reflects the need to re-use this highly used commodity.

Efforts have been made by large industries such as the airline industry where collection of the "one-time use" materials is relatively easy and contained. There has also been an expansion of offerings of several types of disposal containers in public places such as airports, hotels, shopping areas, offices and schools to separate trash from re-usable plastics. The Bag2Bag® recycling by Hilex Poly is successfully recycling plastic grocery bags and their recycling barrels can be found in just about any grocery store nationally.

Other methods of collection are curbside recycling, auto dismantling, community recycling and industrial recycling.

3.4.2.3 Methods used in automotive parts recycling

Methods of generating PIR or PCR recycled materials range from mechanical or manual separation, grinding, chemical or mechanical removal of paints/adhesives, melt filtration & pelletization, de-polymerizaton and re-compounding.

Mechanical (by a machine) or manual (by a human) separation of scrap parts takes a scrapped assembly and separates the various parts so that it can be recycled. Since we've seen that so many different plastics are used to make a vehicle, to make recycling viable, materials must be separated into their respective grades of material. This provides the highest quality material at the optimal cost.

Grinding of parts, sprues and runners, typically referred to as regrind, is one of the most common methods used for injection molded plastics. The materials are put in a shredder or grinder which reduces the plastic to smaller pieces. These can then be either melted and reformed into pellets or added directly back into the injection molding press.

Chemical or mechanical removal of paints and adhesives is done to provide clean feed stream of plastic. Residual paints and adhesives typically are of dissimilar chemistry to the plastic they are used on and must be removed. Otherwise, if it remains, it will be a contaminant and reduce the overall properties of the recycled material.

Melt filtration is another method to remove paints, adhesives and other contaminants from plastic. A filter is placed in an extruder and the polymer melt goes through it prior to be re-pelletized.

In some injection processing it is necessary or ideal to re-melt and re-pelletize the regrind. This may be due to the need to maintain a specific color or a low number of injection shots versus barrel size where the regrind size may be too large to melt efficiently.

De-polymerization is used on thermoset plastics where re-melting is not an option. This method breaks the polymer back down to the monomer level. This can be done by chemical, pressure, moisture, heat or a combination of these conditions. The monomers are then used again to make the original polymer structure. An example of this is in the recycling of headliner foam where the polyurethane foam is de-polymerized and the resulting polyol monomer is re-used to make more polyurethane foam. This type of recycling can have negative environmental impacts where chemical depolymerization is needed or can result in high energy and high cost to effectively depolymerize the polymer.

Re-compounding takes the recycled plastic and adds resins, fillers, stabilizers or other additives to bring the plastic formulation back to the required property performance level. This is the role that many recyclers in the industry play to take PIR and PCR plastics back to the high quality levels required in automotive applications.

3.4.2.4 Legislation supporting recycling

3.4.2.4.1 Existing legislation

The major push for recycling legislation for the automotive industry has been established and led by the European Union. The End of Life Vehicle (ELV) recovery and recycling legislation objective and direction as taken directly from the EU Commission:

> ***End-of-life vehicles***
>
> *Motor vehicles which have come to the end of their useful life and are no longer suitable for use generate millions of tons of waste. To minimize the impact on the environment, to ensure the better reuse of the materials and to improve energy conservation, European Union legislation stipulates how the new vehicles should be designed and how this waste should be collected and treated.*
>
> ***ACT***
>
> *Directive 2000/53/EC of the European Parliament and of the Council of September 18, 2000 on end-of-life vehicles – Commission Statements.*
>
> ***SUMMARY***
>
> ***WHAT DOES THE DIRECTIVE DO?***
>
> *It sets out measures to prevent and limit waste from end-of-life vehicles (ELVs) and their components and ensures that where possible this is reused, recycled or recovered.*
>
> ***KEY POINTS***
>
> - *Vehicle and equipment manufacturers must factor in the dismantling, reuse and recovery of the vehicles when designing and producing their products. They have to ensure that new vehicles are:*
>   - *reusable and/or recyclable to a minimum of 85 % by weight per vehicle*
>   - *reusable and/or recoverable to a minimum of 95 % by weight per vehicle.*
> - *They may not use hazardous substances such as lead, mercury, cadmium and hexavalent chromium.*
> - *Manufacturers, importers and distributors, must provide systems to collect ELVs and, where technically feasible, used parts from repaired passenger cars.*
> - *Owners of ELVs delivered for waste treatment receive a certificate of destruction. This is necessary to deregister the vehicle.*
> - *Producers meet all, or a significant part, of the costs involved in the delivery to a waste treatment center. There is no expense for the vehicle's owner except rare cases where the engine is missing or the ELV is full of waste.*
> - *Waste treatment centers must apply for a permit or register with the competent authorities.*

- *ELVs are first stripped before further treatment takes place. Hazardous materials and components are removed and separated. Attention is given to the potential reuse, recovery or recycling of the waste.*
- *Clear quantified targets for annual reporting to the European Commission exist for the reuse and recovery of ELVs. These have become increasingly more demanding.*
- *EU countries report to the European Commission every 3 years on the implementation of the directive.*
- *The legislation applies to passenger vehicles and small trucks but not to big trucks, vintage vehicles and special use vehicles.*

Separate legislation applies to the reuse, recycling and recovery of vehicle parts and materials.
BACKGROUND
Every year, ELVs create between 8 and 9 million tons of waste in the EU [15].

Other countries have adopted similar legislation. China ELV [16] programs are defined by the End-of-Life Vehicle Recycling Regulations, the Automotive Products Recycling Technology Policy and the Regulation of Remanufacturing of Pilot Automotive Parts. Their approach and targets are similar to the EU ELV. Japan's ELV [17] program stems from lack of space for disposal of scrap materials from vehicle dismantling but does set targets for recycling or recovery targets as does the EU ELV.

Canada has enacted ELV legislation effective September, 2017 [18]. Unlike EU and China, there are no specifics on recycling or recovery. Instead the legislation focuses on the disposal of liquids and control of emissions in the recycling process.

In the United States, there is no specific ELV legislation. The auto industry in the US has taken a very proactive role in automotive recycling more from a business and manufacturing standpoint. Below are a few agencies in the US dealing with automotive recycling:
- *Automotive Recyclers Association: www.a-r-a.org/*
- *The Alliance of Automobile Manufacturers: www.autoalliance.org/*
- *the Institute of Scrap Recycling Industries (ISRI): www.isri.org*
- *Automotive Recyclers of Michigan: www.automotiverecyclers.org/*

In general, legislation is approached by the Environmental Protection Agency (EPA) in terms of pollution control, emissions and management of resources such as water.

3.4.2.4.2 Effectiveness of legislation

In EU countries and Japan, collection and legislation is well established, but acceptance of the use of recycled materials in automotive applications has only come into play in recent years. Oddly enough, in the US and Canada where legislation is recent or non-existent, a sound recycling infrastructure exists and use of recycled materials in automotive components has been used successfully for decades.

3.5 Testing considerations for bio-based or recycled materials

3.5.1 Property variation potential

For any automotive application it is imperative to meet the part performance requirements. Typically, there is a defined set of material properties – a material specification- in addition to the part performance tests. Be it fossil fuel, bio, or recycled based materials, the properties defined in the material specification must be met with statistical data (compiled from minimum 3 to 6 different manufacturing lots and minimum 30 data points per value) where noted.

In the case of bio-based fillers, the statistical spread on the data will be greater than non-bio-based fillers such as glass or mineral. As a product of nature, bio-based properties can and will fluctuate due to changes in climate and soil type, resulting in a wider distribution of data compared to materials that are synthetically manufactured such as glass. Regional effects will play a role in properties as well since climate and soil types are regional. For example, in tropical climates crops will be different than in colder climates.

This does not preclude use of bio-based materials in engineered applications however. With the application of solid statistical data, an understanding of the variability can be applied to properly design components with these types of materials. Also, not all bio-based materials will have a high level of variability. Poly(ethylene) and Poly(trimethylene terephthalate) are two examples of bio-based plastics with narrow property distribution.

3.5.2 Automotive specifications

With the previous considerations on potential increase variability of bio-based/ natural plastics, changes in ranges, test methods and test sample preparation may be necessary. For example, a reinforcement material such as glass will have a very consistent set of properties due to the tighter range of properties in the incoming. For those bio-based materials, specifications may need to include not only reasonable ranges on properties to account for natural material variation, but they also may need to have notations for sources and feed stocks if non-food sourced is a requirement or concern.

3.5.3 Volatile Organic Compounds/Fogging/Mildew/Odor

For the natural fiber or filler additives, concerns over volatile organic compounds (VOCs), odor and mildew are increased due to the inherent nature of these materials. For the VOC concerns, if these types of materials are processed at temperatures above 400°F (195 °C), then breakdown of the structure of these materials will occur causing release of VOCs. Moisture absorption is an issue since water is absorbed and retained. Absorption can be in levels as high as 50 % or more and will cause odor and mildew in certain conditions.

3.5.4 CAE data

Data for computer-aided engineering (CAE) or computer modeling can be somewhat challenging for composite plastics using bio-based materials, particularly those made for lightweight applications using compression molded materials. Typical test specimens used for standard property generation do not readily lend themselves to these types of materials. This is in part due to the material thickness and non-homogenous nature of the composites. Development of test methods and standards needs to be done in the area to obtain more accurate modeling.

3.6 Historical use of sustainable materials in automotive applications

3.6.1 Are sustainable materials new to automotive?

Although it may seem that the term sustainability and use of such materials is a twenty first century concept, use of bio-based plastics goes back to Henry Ford's 1941 Soybean plastic car [19]. Who in automotive circles has not seen the infamous picture of Henry Ford about to hit the trunk of what was supposed to be made of a very impact resistant soybean plastic? It never made it into production, but the idea was there.

Many early polymers were made from natural materials such as cellulose or natural rubber.

In reality, bio-based materials have been used in automotive for decades. These include nylon materials made from naturally derived polyols and cellulose or wood flour filled plastics, such as in Figure 3.2.

3.6.2 Evolution

With all of the new autonomous, electric and highly connected vehicles, recyclers will face new challenges, as illustrated. Electronics recycling will weigh in heavily in this new arena.

With interior trends looking at multi-user vehicles, a potential trend may be easy to install interior components that can be easily removed or added to

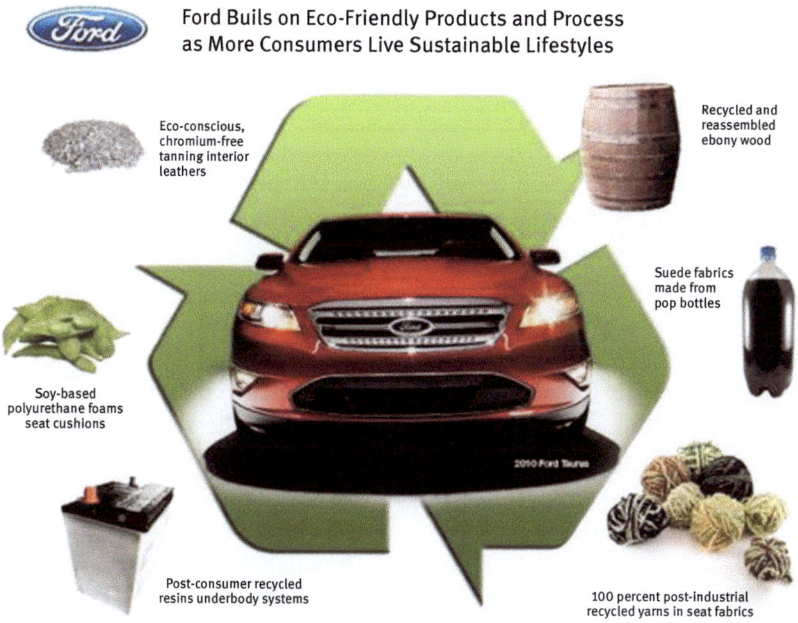

Figure 3.2: Examples of applications. Smart Vehicles Require Smarter Auto Recyclers [20].

provide customization for each user. Alternatively, the interior of a multi-user vehicle may need to be refreshed more often since hours of use per day will increase. This will create a demand for a larger number of interior parts which can be a feed stream for recyclates and a consumer of those same or other recycled materials.

Circular economy is the current talk today and is used in many circles. The definition of a circular economy is [21]:

> Looking beyond the current 'take, make and dispose' extractive industrial model, the circular economy is restorative and regenerative by design. Relying on system-wide innovation, it aims to redefine products and services to design waste out, while minimizing negative impacts. Underpinned by a transition to renewable energy sources, the circular model builds economic, natural and social capital.

It is also defined by seven principles [22]:

(1) **Prioritize Regenerative Resources:** Ensure renewable, reusable, non-toxic resources are utilized as materials and energy in an efficient way.

(2) **Preserve and Extend What's Already Made:** While resources are in-use, maintain, repair and upgrade them to maximize their lifetime and give them a second life through take back strategies when applicable

(3) **Use Waste as a Resource:** Utilize waste streams as a source of secondary resources and recover waste for reuse and recycling.
(4) **Rethink the Business Model:** Consider opportunities to create greater value and align incentives through business models that build on the interaction between products and services.
(5) **Design for the Future:** Account for the systems perspective during the design process, to use the right materials, to design for appropriate lifetime and to design for extended future use.
(6) **Incorporate Digital Technology:** Track and optimize resource use and strengthen connections between supply chain actors through digital, online platforms and technologies that provide insights.
(7) **Collaborate to Create Joint Value:** Work together throughout the supply chain, internally within organizations and with the public sector to increase transparency and create joint value.

This is not new to the automotive industry, however. In 2002 came the release of the book "Cradle to Cradle" written by William McDonough and Michael Braungart. In it they also promoted this type of thinking with the ideals [23]:
(1) A question of design
(2) Why being "less bad" is no good
(3) Eco-effectiveness
(4) Waste equals food
(5) Respect Diversity
(6) Putting Eco-effectiveness into practice

These concepts were heartily embraced by William Clay Ford Jr., who worked to apply them across not only the Ford products, but offices and manufacturing sites [23].

This type of thinking has aided in promoting use of sustainable materials in the automotive industry and will continue to do so moving into the future.

3.7 Bio-based or recycled materials: analysis of opportunities & concern

With all of this great press and global recognition of the need to use renewable sources and recycled products, why do we still have such a wide variety of fossil fuel-based plastics in use?

The current plastic industry has been in place for decades and use of fossil fuels has been a source of raw materials for most of that time. In the 1950–1970's the cost of gas was very low, global consumption was low and no one worried about crude oil as a finite source. As time moved on and prices and demand increased, it was apparent that the raw material feed stream was indeed finite. Concerns for global warming also made harvesting carbon from the ground in the form of crude oil a negative impact on greenhouse gas evolution. Development with renewable sourced materials was fueled by these factors and great strides in the infrastructure have been made, but they still have a way to go before becoming as main stream as their current fossil fuel counter parts.

3.7.1 Concerns with bio-based materials

Some clashes exist in the way the agricultural industry works versus the automotive industry. The expectations with respect to performance and properties of materials in the automotive world are not in line with the variations experienced in the crops, seasons and soil differences.

Low oil prices subdue the efforts put into bio-based products because the overall business case to invest and build the necessary infrastructure is diminished.

Concerns with global hunger and use of crops and or agricultural real estate have come under huge criticism. Are we exchanging car parts for needed food?

Cost is a significant factor in the production of automobiles, and emerging technologies that do not offer cost savings with equivalent performance are hard to adopt. This challenges the development of new technologies not only to offer a sustainable equivalent, but one that either out performs the incumbent (weight savings, quality, properties) or offer a more cost-effective solution.

3.8 Future growth and market direction

Still there are a number of companies which are into full-scale commercial production of renewably sourced polymers[24]:

- **Novamont (Italy)** is one of the leaders and is known for its Mater-Bi starch based resins and the Matrica project
- **Arkema** (France) produces PA11 and PA10
- **Matrica is a joint venture between Novamont (IT) and Versalis (IT)**
- **Yield10 Bioscience** (bought Metabolix) produces **PHA.**
- **Reverdia** is a joint venture between **DSM** (Holland) & **Roquette** (France). They produce succinic acid
- **Solvay** (Belgium) is famous for its brand Epicerol, a biobased glycerol epichlorhydrine for bio-epoxy.
- **Bioamber (Canada)** produces succinic acid.
- **NatureWorks** is a joint venture between Cargill (USA) & PHH (Oil company in Thailand). They produce **PLA.**
- **Total Corbion PLA** is a joint venture between Total (FR) and **Corbion** (NL). Corbion was previously called Purac. They produce PLA and the headquarters are in NL. **Total Corbion joint venture**

- Succinity GmbH is joint venture between **BASF** (Germany) & Corbion (NL). They produce Succinic Acid. The factory is based in Spain.
- **Synvina** is a joint venture between **BASF** (Germany) & **Avantium** (NL). They produce PEF and the factory is based in Antwerp, Belgium.
- Dupont (USA) & Tate and Lyle (UK) produces 1.4 PDO, PTT and PA10.
- **Genomatica** (USA) produces 1.3 BDO, 1.4 BDO and bio-based Caprolactam.
- **Genomatica** supplies 1.4 BDO to **Toray** (Japan) to make PBT.
- **Novozymes** (Denmark) is the global leader for enzyme production.
- Joint venture between **Versalis** (Italy – part of Italian oil company ENI), **Novamont** and **Genomatica** supplies 1.3 BDO. They produce Bio-based Butadiene (bio-BDE). **Versalis to Partner with Genomatica and Novamont for Bio-based Butadiene**
- **Lanzatech** (USA) produces bio-based butadiene (bio-BDE).
- **Verdezyne** (USA) produces bio-based Dodecanedioc acid (DDDA) in Malaysia.
- **PTT MCC Biochem** is a joint venture in Thailand between PTT Public Company Limited and **Mitsubishi Chemical** Corporation.

As the market demand grows for these materials, they will provide excellent, high-performing, cost-effective renewable plastics.

Light weighting is another avenue for renewable materials since the natural fibers offer high strength reinforcement with lower densities. Combined with recycled carbon fibers, a very structural, low density, cost-effective product can be made and be quite suitable for automotive applications.

Although plastics production still out paces recycled material use, it is gaining in momentum both from a performance and cost-effective solution as well. More and more recycled plastics compounding companies have established themselves as producers of high quality plastics and are being used globally in the automotive world.

Perhaps if we referred to recycled materials as experienced plastics, more people would demand that their vehicles be made with such materials!

References

[1] American Chemistry Council. Available at: https://plastics.americanchemistry.com/Automotive/. 2016.
[2] Advanced Plastics are the Future of our Autos. Available at: https://www.plasticsmakeitpossible.com. 16 April 2018.
[3] 13 High Performance Plastics Used in the Automotive Industry. Available at: www.craftechind.com/13high-performance-plastics-used-in-the-automotive-industry/. 2015.
[4] 13 High Performance Plastics Used in the Automotive Industry. Available at: www.craftechind.com/13high-performance-plastics-used-in-the-automotive-industry/. 2015. http://www.craftechind.com/products/.
[5] Available at: www.aftermarketnews.com/ihs-markit-forecast-autonomous-vehicle. Accessed: 12 June 2017.

[6] US Department of Commerce, Bureau of Economic Analysis, 2017 GDP by Industry.
[7] ASTM D6866-18 "Standard Test Methods for Determining the Bio-based Content of Solid, Liquid, and Gaseous Samples Using Radiocarbon Analysis". – This standard is issued under the fixed designation D6866; the number immediately following the designation indicates the year of original adoption or, in the case of revision, the year of last revision. A number in parentheses indicates the year of last reapproval. A superscript epsilon (´) indicates an editorial change since the last revision or reapproval.
[8] United States Department of Agriculture, Bio Preferred program: Available at: https://www.biopreferred.gov/BioPreferred.
[9] EN16785EN 16785-1, Bio-based products – Bio-based content – Part 1: Determination of the bio-based content using the radiocarbon analysis and elemental analysis – EN 16785-2, Bio-based products – Bio-based content – Part 2: Determination of the bio-based content using the material balance method.
[10] ISO 16620-2; This part of ISO 16620 specifies a calculation method for the determination of the bio-based carbon content in monomers, polymers, and plastic materials and products, based on the 14C content measurement. This part of ISO 16620 is applicable to plastic products and plastic materials, polymer resins, monomers, or additives, which are made from bio-based or fossil-based constituents.
[11] Bioplastics News. Available at: https://bioplasticsnews.com/polylactic-acid-or-polylactide-pla/.
[12] E. I. du Pont de Nemours and Company – based in the United States. Also known as Dupont. *Sorona*® is a trademark of Dupont.
[13] ArchDaily article: How Newlight Technologies Creates a Plastic from Air and Greenhouse Gases, Rory Scott. 1 July 2015.
[14] Automotive Recyclers Association Recycling Brochure: Automotive Recycling Industry: Environmentally Friendly, Market Driven, and Sustainable. Published by ARA, the Institute of Scrap Recycling Industries (ISRI) and The Alliance of Automobile Manufacturers. 2016.
[15] European Commission – End of Life Vehicles Legislation; Directive 2000/53/EC.
[16] Ministry of Industry and Information Technology of the Peoples Republic of China. Management Requirements for Hazardous Substances and Recovery Rates on Automobiles June1 2015.
[17] Japan: Law for the Recycling of End-of-Life Vehicles -Automobile Recycling Law. January 2005.
[18] ONTARIO REGULATION 85/16 -UNDER PART II.2 OF THE ACT – END-OF-LIFE VEHICLES.
[19] Soybean Car. Popular Research Topics. Dearborn, Michigan: Benson Ford Research Center, The Henry Ford.
[20] Latham, Andy, News Reports & Industry Views on the Future of Cars, Vehicle Revolution. Smart Vehicles Require Smarter Auto Recycler Automotive Recycling magazine. July-August 2018.
[21] Available at: www.ellenmacarthurfoundation.org/circular-economy.
[22] Available at: www.circle-economy.com/the-7-key-elements-of-the-circular-economy.
[23] McDonough W, Braungart M. Cradle to Cradle. New York: North Point Press; 2002.
[24] Available at: www.bioplasticsnews.com/top-bioplastics-producers/.

Christina Britton and Erik Petrovskis

4 Meijer and carbon management

Abstract: Meijer is a family-owned retailer in the Midwest who pioneered the one-stop-shopping concept. This 85-year-old company is committed to energy, refrigerant, and logistics management to reduce the company's carbon footprint intensity, not only because it makes business sense, but because it is the right thing to do. Meijer operates 248 superstores and grocery stores in six states with associated greenhouse gas emissions from lighting, heating, cooling, refrigeration, and logistics. Through a combination of focused engineering, and proactive lifecycle management throughout its operations, Meijer has been able to reduce the carbon intensity of its operations as the company continues to grow. Meijer's dedicated energy and engineering team is well on the way to reducing electricity use in stores by 20% by 2020, based on a 2010 baseline. Implementing energy efficiency projects to meet this goal, along with proactive management of refrigerant systems, has enabled this team to continue to drive down greenhouse gas emissions intensity in the retail area. Meijer's distribution and logistics teams are also contributing to this reduction through replacement of older tractor trailers with clean diesel and hybrid models in addition to innovations in logistics planning to minimize truck miles traveled. Looking to the future, Meijer sees the next frontier of decarbonizing its operations in proactive lifecycle management, technology, and systems innovation.

Keywords: energy management, carbon emissions, refrigeration

4.1 Introduction

At Meijer, we believe that to be a good company, we must be a good neighbor. We integrate environmental stewardship principles into our daily operations, not only because it makes good business sense, but because it is the right thing to do. Meijer is a Grand Rapids, Michigan-based retailer that operates 248 supercenters and grocery stores throughout Michigan, Ohio, Indiana, Illinois, Kentucky, and Wisconsin. A privately owned and family-operated company since 1934, Meijer pioneered the "one-stop shopping" concept and has evolved through the years to include expanded fresh produce and meat departments, as well as pharmacies, comprehensive apparel departments, pet departments, garden centers, toys and electronics.

Fred Meijer, the longtime leader and owner of the company, once said, "Set your standards – then live by them," and we try to achieve this ideal every day by operating

This article has previously been published in the journal Physical Sciences Reviews. Please cite as: Britton, C., Petrovskis, E. Meijer and carbon management *Physical Sciences Reviews* [Online] 2020, 5. DOI: 10.1515/psr-2019-0022.

https://doi.org/10.1515/9783110597783-004

in a way that fulfills the Meijer Environmental Commitment. One of the core tenants of this commitment is energy, refrigerant, and logistics management to reduce the company's carbon footprint intensity. Climate change is one of the many challenges driving sustainability to the forefront of planning and operations for producers, distributors, and retailers world-wide, and Meijer recognizes its responsibility to reduce its environmental impact, including its greenhouse gas (GHG) emissions. Since the 1970s, Meijer has been working diligently to reduce energy demand through energy efficiency and improved logistics, and since the early 2000s, the company has worked to reduce refrigerant gas leakage. Through focused engineering and proactive lifecycle management, the Meijer engineering, logistics, and distribution teams have made significant strides in reducing carbon emissions from operations; the next frontier of carbon reductions comes from utilizing technology to optimize systems.

4.2 Meijer carbon emissions

Meijer operates 240 stores and 212 gas stations in six states. With nearly 24/7 operations at these retail locations, demand for lighting, equipment, heating, and cooling adds up to a substantial amount of energy use. Additionally, Meijer operates five distribution complexes, five manufacturing facilities, two central fill pharmacies, several corporate offices, and a fleet of over 250 trucks.

Scope 1 GHG emissions result from natural gas usage for heating, diesel fuel use for trucks and generators, refrigerant gas leakages, and gasoline use for Meijer-owned vehicles. Scope 1 emissions account for 27% of total GHG emissions.

Scope 2 GHG gas emissions come principally from purchased electricity and account for 73% of total emissions. See Figure 4.1 and Figure 4.2 for a breakdown of GHG emissions by scope and by source.

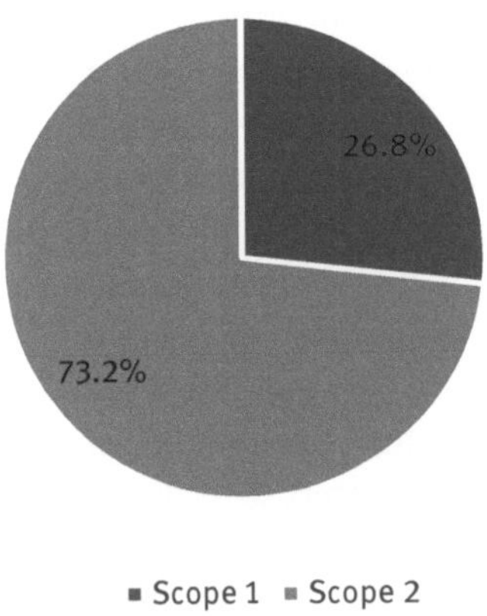

Figure 4.1: Total GHG emissions by scope*. *Total GHG emissions include scopes 1 and 2.

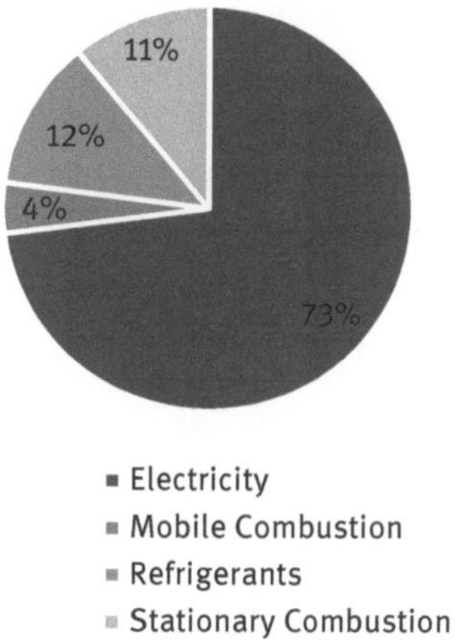

Figure 4.2: Total GHG emissions by source*. *Total GHG emissions include scopes 1 and 2.

Scope 3 emissions have not yet been quantified, but it is important to the company to work towards characterizing, quantifying, and managing scope 3 emissions.

4.3 Energy efficiency

From the day that Meijer opened the doors of the first grocery store in Greenville, Michigan in 1934, the business has operated under a culture of thrift. Opening a new business during the great depression, this culture was essential to the early success of the company. This culture of thrift has enabled Meijer to grow responsibly, even during economically challenging times, and it is one of the key drivers of energy efficiency initiatives. In the early 1970s, the energy crisis in the United States spurred the company to put in place its first policies and procedures to reduce unnecessary energy use, and many of the practices put in place almost 50 years ago are still adhered to today. A 1974 employee newsletter from the Meijer archives includes practices such putting in place lighting efficiency measures, building full tractor-trailers from the distribution centers, and turning off equipment when not in use – all practices that the company still uses today.

Meijer has been proactive in evaluating how emerging technologies can be used within our operations to reduce electricity and natural gas usage both within new stores and as retrofit projects. Starting in the early 2000s, Meijer convened an in-house energy and engineering team wholly dedicated to store performance and evaluating and implementing energy efficiency opportunities in our stores and gas stations, including LED lighting, HVAC replacements, VFD installations, occupancy sensors, closed refrigerated cases, and more.

The reason that this team is focused on retail operations is that over 85% of the company's GHG emissions come from our retail operations and that energy efficiency projects determined to meet a certain level of return can be replicated throughout the store fleet due to the similarity of building specifications and operations. See Figure 4.3 for a breakdown of Meijer GHG emissions by business area.

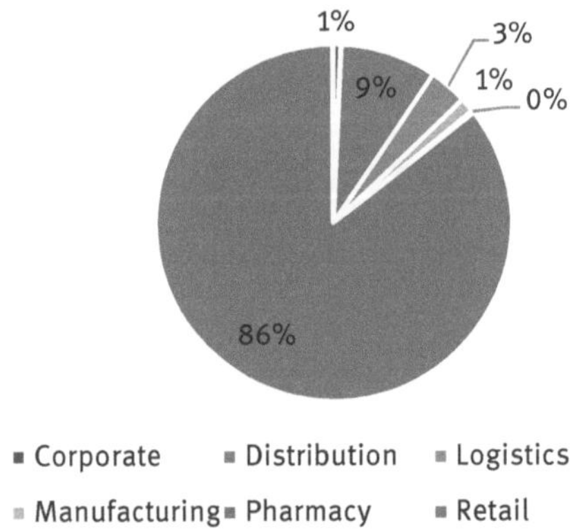

Figure 4.3: GHG emissions by business area.

Understanding that a common vision drives action, the Meijer engineering team set an ambitious energy savings goal in 2010 to reduce 20% of electricity in stores by the year 2020 and is on target to meet it though creativity and innovative engineering practice.

4.3.1 Key energy efficiency tactics

As Meijer's energy efficiency program has developed over the years, a few important tactics have emerged.

1. Meijer designs new stores to meet LEED certification criteria, and has been externally certifying one store annually since 2007, resulting in 12 LEED Certified stores. When remodeling a store, we look to utilize LEED design elements as part of the project.
2. The engineering team uses EnergyStar to benchmark stores annually, which allows for an objective view of building energy performance and an annual snapshot of where our biggest opportunities remain.
3. When evaluating an energy savings technology, we pilot the project in one or more stores to measure the performance and to validate the results. Once the technology has been validated, we develop a plan to implement across our fleet of stores.

4. Meijer has been performing energy efficiency projects for numerous years. As the program has developed, the engineering team has focused on combining projects into the same stores each year, to maximize the results, while minimizing the customer impact.
5. The engineering team works hard to align priorities between all groups and maximize benefits in addition to energy savings. For example, when replacing open freezer cases with closed cases, more product can be stored in the same amount of space; this leads to less stocking labor, which saves the company money. In addition, with closed cases, the aisles are warmer, which makes customers more comfortable, leading to greater satisfaction and shopping times.
6. Proper equipment maintenance is essential to maximizing energy savings and minimizing lifetime emissions.

4.3.2 Awards and recognition

Recently, Meijer's Bay City location placed second in the Michigan Battle of the Buildings Retail Category by reducing energy consumption by an impressive 20.57% over the 2017 calendar year. This project included an update of many of the store's mechanical systems including rooftop HVAC units and refrigerated and frozen cases, using key LEED design principles.

4.3.3 Energy efficiency project examples

Project Example 1: The engineering team needed to change out about half of the cases at a store, and the other half of the cases were new. The team performed an engineering analysis that determined that if they changed out half of the cases, they would have had to add an additional refrigeration compressor, which would have meant more refrigeration gas and more energy use. The team went back to the drawing board to look at the system in a more holistic way. What would they have done if they were to redesign the system from scratch? The team determined that if they took the whole refrigeration lineup out and put a whole new lineup in with closed cases, it would save money and energy. This eliminated the need for an additional compressor and more energy and refrigerant gas. They were able to reuse half of the cases in a different store, which added to the project savings. This is a good example of how the engineering team uses lifecycle management and sound engineering practice to find creative energy- and cost-saving solutions.

Project example 2: As LED technology matured in the early 2000s, the Meijer engineering team recognized the potential energy and cost savings of retrofitting lighting to utilize this technology. Starting in about 2013, Meijer began implementing large retrofit projects throughout the store fleet. With thousands of light fixtures at each retail location, doing a total changeover to LED was no small feat. It is predicted that the total fleet changeout will be completed in about 4 years, and this is a win for Meijer in several ways. Not only does it directly save electricity, but LEDs also have a longer lifecycle, meaning reduced cost for replacement and less waste. Now, the lighting that we put in the freezers and coolers will last about 18 years with an occupancy sensor.

Looking towards the future, the engineering team is now focused on engineering systems in a more data-based and detailed way since many of the capital projects with direct energy savings potential have already been implemented. The focus now lies in lifecycle management, software, and details. One example is fine-tuning the programing of older HVAC systems to run more efficiently. Better HVAC performance means that the stores are drier so there is less condensate on the glass doors of the refrigerated and frozen cases. This also means that those cases work less and use less energy – all without adding more equipment.

4.4 Refrigeration

Operating 240 superstores with significant refrigerator and freezer capacity at each means that Meijer manages a large amount of compressed refrigerant gas. Some refrigerant gases have a global warming potential (GWP) of more than 10,000 times that of carbon dioxide. Proactively and aggressively managing our refrigerant gas systems to minimize leakages is important to Meijer to reduce our GHG emissions, reduce costs, and maintain regulatory compliance. Refrigerant gas emissions account for 12% of Meijer total GHG emissions.

4.4.1 Refrigerant management performance

While Meijer has been actively managing refrigerant gases for many years in order to manage costs, in 2012, the company saw the value in joining the U.S. Environmental Protection Agency's GreenChill Partnership to better benchmark, track, and manage refrigerant gases in our operations. GreenChill is an EPA Partnership with food retailers to reduce refrigerant emissions and decrease their impact on the ozone layer and climate change. Since joining the program, Meijer has reduced total refrigerant gases leaked by 18%. The company achieved this absolute reduction while adding 20% more retail square footage over the same time period (42 new stores). The first year that Meijer submitted data to GreenChill, we had an 11% refrigerant leak rate, which was 2.1% better than the GreenChill Partner average. In 2017, we achieved an 8.5% leak rate, which makes Meijer a top performer in the GreenChill program. Meijer has achieved the Superior Goal

Achievement Award from the GreenChill Partnership three times since joining the program in 2012, and each year, we set a more aggressive leak rate goal to help drive our efforts in this area. Meijer's annual GreenChill goal and leak rate since joining the program are presented in Figure 4.4.

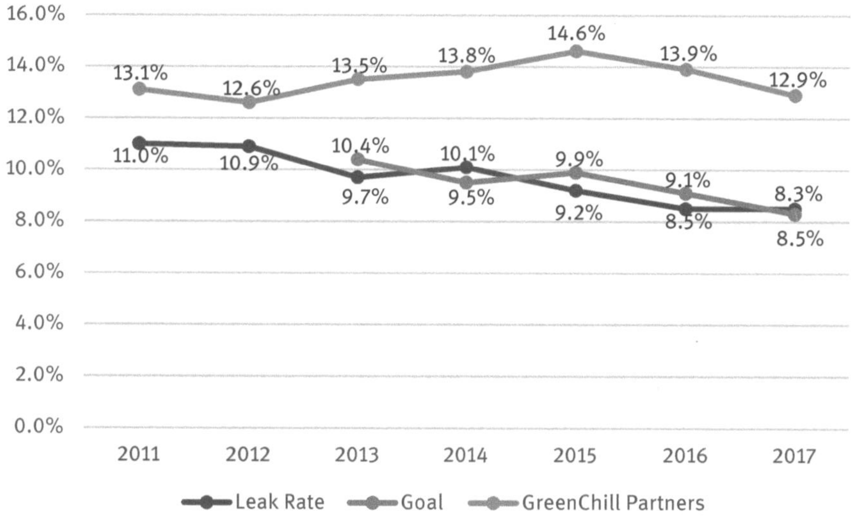

Figure 4.4: Meijer refrigerant leak rate and GreenChill annual goal.

Refrigerant gases that are commonly used for retail refrigeration have a GWP between 1 and 14,400, and the average for the GreenChill partnership is 2,962. Meijer's average GWP of installed refrigerant gas in 2017 was 3,527. Despite having a higher average GWP, Meijer is very effective at managing refrigerant leaks and has been recognized as a top performer in the GreenChill Partnership.

The main reason that Meijer has a higher GWP than the average for GreenChill is, counterintuitively, because Meijer phased out R-22 faster than was required by federal regulation. R-22 has a high ozone depleting potential, which lead the U.S. and other countries to pass laws to phase out the manufacture and use of the gas. Meijer decided to be proactive and phase out the harmful gas several years before it was required by law. The best alternative at the time was R-404A, which has a GWP of 3,920, much higher than R-22, which has a GWP of 1,810. Meijer completely phased out R-22 well ahead of the 2020 regulatory deadline. Since Meijer completed the changeouts, a new refrigerant gas has come on the market called R-448A, which has a GWP of 1,390. All new stores and refrigeration systems are charged with R-448A, but it is costly to do yet another gas changeout, just a few years after we retrofitted the entire retail refrigeration fleet. The engineering team has piloted this retrofit to better understand what the change will cost and continues to review options to remove R-404A.

4.4.2 Key refrigerant management tactics

Since we are locked into such a high GWP of our installed refrigerant gas for the time being, it becomes even more important for us to quickly detect and fix refrigerant leaks as they happen. Some key tactics have been developed to most effectively manage these systems.

1. The engineering team installs remote monitor leak detectors on the refrigeration systems. Detection points are installed in strategic locations, like behind cases and near walk-ins; this allows for the quick detection and response of a leaking system to minimize the impacts.
2. On a quarterly basis, all refrigeration systems are inspected for leaks.
3. Refrigerant technicians do a full leak check of any unit that has to be charged with additional refrigerant, even when it is not required by law.
4. In new and remodeled systems, the engineering team designs new refrigeration systems to use as little refrigerant as possible. Lower refrigerant charge means less possible emissions. For example, compared to the refrigeration system in a 2005 Meijer store, the systems installed today use a distributed refrigeration system that results in a 35% reduction in refrigerant gas charge and a 41% decrease in emissions, all with a 31% increase in load per store and no increase in energy use (Figure 4.5).

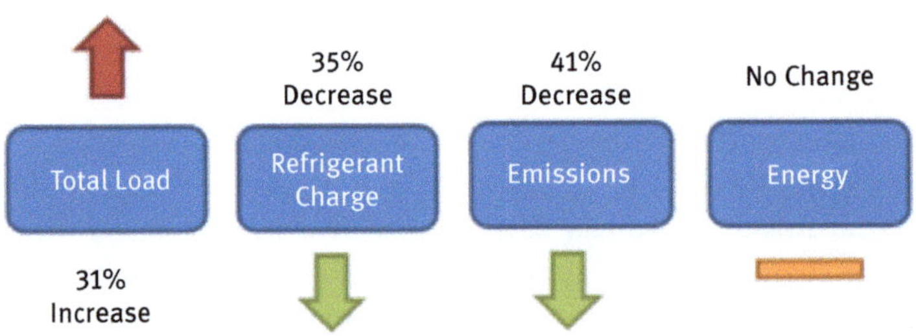

Figure 4.5: 2005 vs. 2018 refrigeration systems.

Looking towards the future, another alternative for Meijer and other retailers is advanced refrigeration systems that use significantly less refrigerant gas for the same amount of cooling as a traditional system; or these systems rely on low-GWP gases such as CO_2. The Meijer engineering team has begun to actively evaluate these options for future refrigeration systems.

4.5 Distribution

4.5.1 Logistics

Meijer operates a fleet of approximately 250 semi-trucks that, collectively, cover an estimated 25 million miles each year. Utilizing innovative technologies to improve fuel efficiency, this fleet is one of most environmentally friendly in the country. In 2009 Meijer partnered with Daimler Freightliner to be the first fleet in North America to operate ten Proto-type 2010 EPA Emission Freightliner Cascadia's. By 2014 Meijer was the First Fleet in North America to have 100% of our Private fleet of 225 trucks meeting 2010/2014 EPA Emission and SmartWay Certification, resulting in a fleet carbon intensity reduction of nearly 60%.

We achieved this goal with a new fleet of "clean diesel" vehicles that utilize selective catalytic reduction (SCR) technology. This technology injects a liquid-reductant agent, known as Urea or Diesel Exhaust Fluid (DEF), through a special catalyst into the exhaust stream of the truck's diesel engine. The DEF breaks down the polluting elements, converting nitrogen oxides into nitrogen, water and small amounts of carbon dioxide (CO_2), which is then expelled through the vehicle tailpipe. The technology has produced a dramatic improvement. In fact, 47 of the new 2014 clean diesel semi-trucks have the same emissions as just one of the trucks it replaced, while our fleet's total fuel economy is about 20 percent more fuel efficient than the national industry average.

In 2016, the Logistics team acquired new hybrid models of refrigerated trailers to further reduce emissions and put in place a protocol to purchase hybrid models for all new refrigerated trailer acquisitions. Once parked at the store or warehouse, they can be switched from the diesel engine over to electric using power from the buildings. This reduces operating noise, as well as pollution from the diesel fuel.

The Meijer Fleet has been a partner in the U.S. Environmental Protection Agency SmartWay program since 2005, and this partnership has been invaluable to the continued improvements in the sustainability of our fleet by helping us to measure, benchmark, and improving freight transportation efficiency. We believe that this program is so valuable that we require all Meijer carriers to be 100% SmartWay certified.

In 2017, Meijer was one of the few carriers out of more than 3,500 North American SmartWay partners that won the 2017 SmartWay Excellence Award. The annual SmartWay Excellence Award honors top shipping and logistics companies for superior environmental performance and additional actions to reduce freight emissions through effective collaboration, operational practices, a robust system for validating and reporting their SmartWay data, and communications and public outreach. SmartWay Excellence Awardees lead their industries in improving freight efficiency and contributing to cleaner air within their supply chains. In 2016, Meijer also won the 50 Top Green Fleets Award from HD Truck Magazine.

As with energy efficiency and refrigerant gas management, the next opportunities for emission reductions come from improving or optimizing systems. Now that the Meier trucking fleet is so efficient, the biggest opportunities are related to technology. The logistics team is working to use technology solutions to optimize routes and minimize vehicle miles traveled. Analytics and big data already help this team become more efficient, and they are starting to investigate the potential of machine learning to further improve their capabilities to improve truck routes.

Additionally, the logistics team is actively investigating electric and alternative fuel trucks, working with several manufacturers as they develop these next-generation technologies. In November 2017, Meijer made a deposit to test out four of the first-generation Tesla semi-trucks that will be released in 2019.

4.5.2 Distribution center efficiency

Operating our distribution centers accounts for almost 10% of Meijer's GHG emissions. As most of the distribution center facilities were built before viable LED technology was available, all require a retrofit project to transition to LEDs. These projects are ongoing, as there is a significant amount of lighting to replace.

Increasingly, Meijer distribution is utilizing automation solutions to increase the overall efficiency of operations. While there is an increase in electricity demand for automation, it is done with the latest efficient technology, and it often allows us to retire some older legacy equipment or buildings that were significantly less efficient. With automation, we can also reduce the building footprint required to process the same amount of product. This, in turn, reduces the amount of space we need to light and air condition. In addition, a lot of our new automated systems are allowing us to build more effective and fuller trailers. The better and fuller we can fill trailers, the less miles we can put on those trailers, which results in a significant reduction in energy usage by the supply chain.

4.6 Where do we go from here?

Starting a grocery business during the Great Depression meant that operating as efficiently as possible was a business necessity. This culture of thrift has permeated the company and helped it grow responsibly, even during challenging times. In the last ten years, reducing the company's carbon footprint intensity has become an additional driver, pushing us towards greater efficiency. Clearly, large strides have been made in constructing a company that runs efficiently, which means minimizing GHG emissions from operations, but what next?

Some of the key next steps for Meijer are shifting the focus from individual operational efficiency programs to a broader and more holistic perspective, both operationally and in the context of sustainability. How can Meijer start to look at carbon management from a more strategic perspective in the context of global climate change? How can we partner with other organizations to share knowledge and work towards a common goal of sustainability?

When taking this next step, understanding the complexities of the broader sustainability context and the nexus between carbon emissions and other areas of sustainability will be essential. It is no longer as easy as replacing a CFL bulb with an LED to get the biggest carbon reduction bang for our buck. For example, if we embrace electric vehicle charging and begin to offer at our stores and gas stations, we will dramatically increase our scope 2 carbon emissions, but that will enable a global reduction of greenhouse emissions if the electric cars are replacing gas-powered vehicles on the road. Another example is our backhauling program for recyclables. By backhauling those materials and aggregating at our distribution centers, we are increasing the amount of diesel fuel burned in our trucks, hence increasing our scope 1 emissions. But in the global context, we are likely reducing total GHG emissions because we are enabling the material to be most efficiently transported through maximized loads directly to the recycling facility. In addition, we are improving our recycling practices to enable the highest and best recycling option for each material through the most efficient process which leads to less virgin materials used in manufacturing, resulting in lower emissions.

To take the next step into carbon management, Meijer will pursue work in several areas.

1. Meijer is working towards characterizing the company's Scope 3 emissions, which is essential to understanding Meijer's carbon footprint in a comprehensive manner.
2. Meijer continues to reduce scope 1 and 2 emissions per square foot, through additional energy efficiency projects. In addition, the engineering team is evaluating digital solutions – utilizing technology and software to fine-tune and optimize systems.
3. Meijer is also seeking ways to evaluate our carbon footprint in a more standardized way, including benchmarking against other retailers through the Retail Industry Leaders Association, and responding to the CDP Climate Change Questionnaire.
4. To reduce scope 2 emissions and support the transition of our electricity grid away from fossil fuels, Meijer is investigating sourcing renewable energy.

Fred Meijer once said that, "If the goal is right, we will find a way to get there." Meijer understands that the right goal is to reduce our carbon footprint intensity, and we will continue to work to find the right way to get there.

Acknowledgements: Thank you to the following Meijer team members for contributing data and operational insights for the important work described herein: Greg Serkaian, Jason Flanigan, Vik Srinivasan, David Hoover, and Dan Scherer.

Susan C. Walkinshaw

5 Applying green chemistry approaches to EPA standard method of analysis for dioxins

Abstract: Green Chemistry principles can be applied while following approved Environmental Protection Agency methods at an accredited lab. The decision-making process involved in selecting a method for the analysis of dioxins in solid matrices while considering green chemistry, method performance and overall cost are presented.

Keywords: EPA Method 1613b, Dioxin, SW-846 Method 3545A, SW-846 Method 3546, SW-846 Method 3550C

5.1 Overview

During the decision-making process for selecting a method for the analysis of dioxins in solid matrices, an accredited lab gives consideration to green chemistry, method performance and overall cost. It can be difficult to make an analytical process greener when the methodology used is prescribed, such as in the case of an Environmental Protection Agency (EPA) method, when the laboratory using the method is accredited by various states and agencies for these analyses and their operations must remain in compliance. Many of the older EPA methods do not have any provision for alteration of the method. Some of the newer methods that are performance based allow for changes or improvements to the method as long as certain performance criteria are met. These changes can also be made in order to make the method greener.

An example of a method that is performance based is EPA Method 1613: Tetra-Through Octa-Chlorinated Dioxins and Furans by Isotope Dilution HRGC/HRMS [1]. As described in the introduction and Scope and Application sections of the method:

> Method 1613 was developed by the United States Environmental Protection Agency's Office of Science and Technology for isomer-specific determination of the 2,3,7,8-substituted, tetra through octa-chlorinated, dibenzo-p-dioxins and dibenzofurans in aqueous, solid, and tissue matrices by isotope dilution, high resolution capillary column gas chromatography (HRGC)/ high resolution mass spectrometry (HRMS).
>
> This method was developed by the Engineering and Analysis division within the USEPA's Office of Science and Technology. The method is used for isomer specific determination to detect the Tetra- through octa- chlorinated dibenzo-p-dioxins and dibenzofurans associated with the Clean

This article has previously been published in the journal Physical Sciences Reviews. Please cite as: Walkinshaw, S. C. Applying green chemistry approaches to EPA standard method of analysis for dioxins *Physical Sciences Reviews* [Online] 2020, 5. DOI: 10.1515/psr-2019-0023.

https://doi.org/10.1515/9783110597783-005

> Water Act (CWA, as amended 1987); the Resource Conservation and Recovery Act (RCRA, as amended 1986); the Comprehensive Environmental Response, the Compensation and Liability Act (as amended in 1986); and the Safe Drinking Water Act and other dioxin and furan compounds amenable to this method. The dioxins and furans may be determined in water, soil, sediment, sludge, tissue, and other matrices using this method. The method is based on EPA, industry, and academic methods.

Currently, at the NSF International Ann Arbor Laboratory water samples are analyzed for dioxins and furans using this method. Investigations are underway to determine what is involved in extending this method to soil and solid samples. At first glance the differences are obvious in that an entirely new extraction procedure is required. The extraction for aqueous samples is a solid phase extraction using a disk. The extraction method for soil and solids in EPA Method 1613b is a Soxhlet extraction with a Dean-Stark trap. A review of this extraction from both a green chemistry and a business perspective shows that it requires considerable laboratory space, hood space, active analyst and waiting time, and fairly significant solvent volumes which adds up to relatively high cost. Since the method is performance based there are other options to be considered. The most straightforward methods to consider as substitutions are the other solid matrix extractions already approved under the Solid Waste program by the EPA, for example; SW-846 Methods 3545A Pressurized Fluid Extraction (PFE) [2], 3546 Microwave Extraction (ME) [3], and 3550C Ultrasonic Extraction [4]. These are the simplest because they are generally acceptable to state accreditation bodies since all are already acceptable to the EPA.

5.2 Method selection

The descriptions of scope and application for each method according to the SW-846 Manual are as follows;

> Method 3545 is a procedure for extracting water insoluble or slightly water soluble organic compounds from soils, clays, sediments, sludges, and waste solids. The method uses elevated temperature (100–180 °C) and pressure (1500–2000 psi) to achieve analyte recoveries equivalent to those from Soxhlet extraction, using less solvent and taking significantly less time than the Soxhlet procedure.
>
> Method 3546 is a procedure for extracting water insoluble or slightly water soluble organic compounds from soils, clays, sediments, sludges, and solid wastes. Its procedure uses microwave energy to produce elevated temperature and pressure conditions (i. e. 100–115 °C and 50–175 psi) in a closed vessel containing the sample and organic solvent(s) to achieve analyte recoveries equivalent to those from Soxhlet extraction (Method 3540), using less solvent and taking significantly less time than the Soxhlet procedure.

In Method 3550C it says, "This method describes a procedure for extracting nonvolatile and semivolatile organic compounds from solids such as soils, sludges, and wastes. The ultrasonic process ensures intimate contact of the sample matrix with the extraction solvent." It also goes on to describe performing the method differently based on the expected concentration of the analytes present and that all instructions need to be followed closely as the performance of the method is limited by the contact time with the sonication probe therefore making it possibly less rigorous than other extractions. This is problematic in that unknown samples are exactly that, unknown; making it unlikely the analyst would know which variation of the method to perform. Also a less rigorous extraction that may not compare favorably to the Soxhlet method is not desirable from a performance standpoint. It may also be subject to varying experience levels and training of the analysts performing the procedure. These alternative methods offer varying advantages and disadvantages with two of them specifically mentioning the reduction of solvent volume and time in relation to the Soxhlet procedure in the application of the method.

After a review of the existing applicable methods, the primary negative of Method 3550C Ultrasonic Extraction for this application is that it is not specified in the SW-846 analytical analysis Methods 8280B (POLYCHLORINATED DIBENZO-p-DIOXINS [PCDDs] AND POLYCHLORINATED DIBENZOFURANS [PCDFs] BY HIGH-RESOLUTION GAS CHROMATOGRAPHY/LOW-RESOLUTION MASS SPECTROMETRY [HRGC/LRMS]) [5] or Method 8290A (POLYCHLORINATED DIBENZO-p-DIOXINS [PCDDs] AND POLYCHLORINATED DIBENZOFURANS [PCDFs] BY HIGH-RESOLUTION GAS CHROMATOGRAPHY/HIGHRESOLUTION MASS SPECTROMETRY [HRGC/HRMS]) [6] as an acceptable extraction for dioxins. That in combination with the other drawbacks has eliminated it from our consideration as a possible alternative to Soxhlet for dioxins.

That leaves Microwave Extraction and Pressurized Fluid Extraction as the possible acceptable alternatives to the Soxhlet Extraction. There are three areas that will be focused on in evaluating these methods to determine which makes the most sense from the green chemistry, analytical chemistry and economic perspectives. These areas are the solvent or solvent system used, the laboratory hood space required and the overall time of the extraction. All of these are directly related to and have a large impact on the overall cost. So by trying to select a greener extraction method there may be cost savings as well. An additional consideration in an accredited lab is the interpretation of the method by the accrediting and auditing bodies. The recommendations from auditors can impact the operations of analytical operations in ways not fully anticipated. The best way to handle this is to realize that the method chosen will in the end be subject to audit and try to consider any issues upfront based on previous audit experience.

5.3 Sample pre-preparation

All of the method choices require some preparation of the samples before the extraction begins, which can include grinding to reduce particle size, separation of liquid, addition of drying agents or determination of percent solids if the results are to be reported on a dry weight basis. The grinding to reduce particle size of the sample seems to be required in Method 1613B; however, in both the newer extraction methods, Microwave Extraction and Pressurized fluid Extraction, grinding dioxin samples is advised against with the statement: “WARNING: The drying and grinding of samples containing PCDDs/PCDFs is *not* recommended, due to safety concerns.” A decision needs to be made on whether this step is necessary to the performance of the method. If not, the safety is greatly improved for the technician performing the extraction as well as saving the time and space involved in this part of the procedure. Obtaining the necessary quality control data without the grinding step and comparing to the performance requirements of the method would be the best way to make this decision regardless of which extraction method is used. Since all of the methods require sample preparation steps, they are not central to the choice of which extraction method to use and will not be considered in making this decision.

5.4 Solvent considerations

The following should be considered when reviewing the solvent usage associated with these extraction methods: the grade of solvent required, the choice of solvent or solvent system used for the extraction, the volume of solvent needed not just during the extraction step but for pre-cleaning the glassware or other lab equipment used in the process, the properties and safety hazards of the solvent chosen, and the cost of the solvent used.

Often for soil matrices a mixed solvent system is used, consisting of a polar solvent to help penetrate any water layer that may be on the surface of the particles and a nonpolar solvent that the analytes of interest are more soluble in. A typical solvent system of this type for dioxin analysis is hexane/acetone. All three of the methods being evaluated call for pesticide grade regardless of the solvent. The microwave extraction method specifies hexane/acetone and the pressurized fluid extraction method specifies toluene as solvent for the extraction with the statement that any solvent can be used based on the performance of the method. The Soxhlet extraction method leaves the choice of solvent up to the analyst as long as the performance of the method is met. The solvents listed in method 1613b include: acetone, toluene, cyclohexane, hexane, methanol and methylene chloride. The NSF International laboratory has used toluene and methylene chloride for the extraction of liquid samples using method 1613b. The switch to methylene chloride from toluene was made due to observations related to cross-contamination of other samples being extracted in the same lab at the same time. This was wasting considerable resources, time and money in re-extracting the other samples. This highlights the need to be aware of the potential for cross-contamination within the laboratory to processes unrelated to the one under consideration. As always, the performance of the various solvents needs to be determined to meet the applicable method requirements.

One of the other considerations is the solvent properties that impact the length of time it takes to concentrate the sample. Another part of the method is the potential clean-up steps that may be needed depending on the matrix interferences present. The clean-ups can require significant additional solvents and steps. Since all three of the methods under consideration for the extraction step will require similar or the same clean-ups for the samples, the solvent needs for clean-up are not being used in this process to decide on the extraction method. All of the methods end with the sample in nonane for injection into the instrument so a solvent exchange is performed during the final concentration step.

For the three extraction methods being considered there is a large difference in solvent volume required, see Table 5.1. The Soxhlet method requires a large volume of solvent to pre-clean the glassware prior to the sample extraction as well as a large volume for the extraction itself. Both the pressurized fluid extraction and microwave extraction use far less solvent, anywhere from 4 to 30 times less. This makes both of these methods preferable to the Soxhlet method based on solvent volume used.

Table 5.1: Solvent volume required.

Method	Solvent volume (L) per 20 samples
1613B Soxhlet	~15
3545A PFE	0.2–4
3546 ME	0.5–1

The safety of the various solvents being considered can be evaluated by comparing the Global Harmonization System (GHS) Hazard Identification information from Section 5.2 of the respective Safety Data Sheets for each solvent. See Table 5.2. Evaluation of the hazards of the various solvents that can be used, especially if a combination is used, leads to most of the hazards being similar. One difference worth noting is that methylene chloride is not flammable which could be an important consideration when looking at the Soxhlet method which requires unattended overnight operation. The other possible solvents are all flammable. The other key difference is that methylene chloride is a potential carcinogen whereas toluene and hexane both show reproductive toxicity. The carcinogenic classification for methylene chloride according to the International Agency for the Research on Carcinogens (IARC) [12] is 2A – Probably carcinogenic to humans. The reproductive hazards of most chemicals do not have any human studies designed to study the effects of long-term low level exposure and the data is limited and based mostly on exposure during manufacturing from the early

Table 5.2: Summary of Section 5.2: Hazards identification from the safety data sheet for the listed solvents.

	Methylene Chloride [7]	Toluene [8]	Hexane [9]	Acetone [10]	Methanol [11]
Flammable liquids	–	2	2	2	2
Acute Oral toxicity	–	–	–	–	3
Acute Dermal toxicity	–	–	–	–	3
Acute Inhalation toxicity - Vapors	–	–	–	–	3
Skin Corrosion/irritation	2	2	2	–	–
Serious Eye Damage/ Eye Irritation	2	2	2	2	–
Reproductive Toxicity	–	2	2	–	–
Carcinogenicity	1B				
Specific target organ toxicity (single exposure)					
Target Organs -					
– Central nervous system (CNS)	3	3	–	3	–
– Respiratory system, Central nervous system (CNS)	–	–	3	–	–
– Optic Nerve	–	–	–	–	1
Specific target organ toxicity (repeated exposure)					
Target Organs -					
– Liver, Heart, Blood	2	–	1	–	–
– Kidney, Liver, spleen, Blood	–	2	–	2	1
Aspiration Toxicity	–	1	1	–	–

twentieth century before modern safety precautions were implemented. The data collected for toluene [13] showed no association between exposure and male fertility; however, a significant association was found between exposure to toluene and female infertility. All of the hexane [14] data are in animal studies based on oral and inhalation routes. In the analysis laboratory, oral exposure is not an expected occurrence. The inhalation data are more relevant, but the correlation of the data to human exposure is more complicated. Also, the hazards are only one part of the evaluation process; the other is a risk assessment. The Occupational Safety and Health Organization (OSHA) [15] clearly defines the difference in these two factors.

> Hazard classification does not involve an estimation of risk. The difference between the terms hazard and risk is often poorly understood. Hazard refers to an inherent property of a substance that is capable of causing an adverse effect. Risk, on the other hand, refers to the probability that an adverse effect will occur with specific exposure conditions. Thus, a chemical will present the same hazard in all situations due to its innate chemical or physical properties and its actions on cells and tissues. However, considerable differences may exist in the risk

posed by a chemical, depending on how the chemical is contained or handled, personal protective measures used, and other conditions that result in or limit exposure.

As a full risk assessment is not possible here, each laboratory will have to evaluate the handling conditions in their own area. These play a large part of the safety of using a particular solvent or solvent system, so the choice is dependent on the laboratory it will be handled in and the training of the analysts and technicians performing the work. In the Global Harmonized System (GHS) of classification of hazards the lower the rating, the higher the relative risk.

The last consideration is the cost of the various solvents. The total cost is more than just the initial purchase price; it should include the disposal cost of the solvent waste as well. Toluene, acetone and methylene chloride are all listed as hazardous waste so the disposal costs of these are all similar. Hexane is the only one not a listed waste, however, it is generally used with acetone which is listed so there is no cost savings from a waste perspective to the hexane/acetone solvent system. The costs are estimated by calculating the per liter cost based on purchasing a case of four liter bottles of the appropriate pesticide grade solvent from a scientific supply warehouse. Toluene and hexane have an equivalent cost with methylene chloride costing an additional 20 % and acetone 30 % more.

Based on the specific situation in the NSF International laboratory regarding the other analyses performed, the first choice would be to use methylene chloride. It is purchased in large bulk tanks at the required purity and the safety measures for handling it are already in place. The proper disposal measures for this waste stream are also in place. Again this choice is specific to each laboratory based on what other work is being performed, the solvents and solvent grades already used, the prices negotiated with suppliers and preferred vendors, the waste generated and the safety systems for handling an existing versus new solvent. All choices would have to be verified relative to the specifications of the applicable analysis method to make sure it meets all performance requirements. This does lead to the "best" choice being different in various laboratory settings.

5.5 Fume hood space

Fume hood space is an important, if sometimes overlooked, part of planning a new analysis. New fume hoods can cost approximately $40,000 on average based on initial installation and one-time set up costs. The continued operation and maintenance costs are between $7,000 and $12,000 per year over the lifetime of the hood depending on the type of hood [16]. This can be a substantial cost for a smaller laboratory. A side note to hood usage also includes the normal operations and the positioning of the sash when the hood is in active use by a scientist compared to when no scientist activity is taking place. Consciously and consistently closing the hood sash can result in large cost savings as evidenced at Harvard University with their Shut the Sash Program, where energy costs were reduced by 70 % [17, 18]. This is true of a variable air volume hood when the hood is not actively being used; however a constant air volume hood draws the same amount of air volume regardless of hood sash position.

In examining the three extraction methods under consideration for the hood space requirements, there is a large difference. The Soxhlet extraction requires a fume hood for the entire 16 h or more extraction period. In order to extract a reasonable number of samples in a set, for example 20, three 6 foot fume hoods would be needed for the glassware set-up used. In addition to the hood space required, a source of cold water to run through the condensers is needed for the entire extraction time. Also each sample set-up needs an adjustable heating mantle or heating plate. See Figure 5.1 for Soxhlet extraction apparatus [1]. In contrast to this, the microwave unit used for the microwave extraction can be placed on the countertop with an exhaust line to a vent duct. The only hood space required would be to prepare and load the samples into the extraction vessels, and remove the extracts after the extraction is done. This may or may not require a dedicated hood, depending on the volume of work already performed in existing hoods within the laboratory. At most, the microwave extraction would require a single hood. The unit that performs the pressurized fluid extraction can also be placed on the countertop and it does not need to be vented. The same hood requirements would apply as to loading the samples into the extraction vessels as with the microwave extraction approach. The microwave extraction and the pressurized fluid extraction are closed systems which allow them to be used on the countertop. Both the microwave and pressurized fluid extractions need a single hood for only a portion of the time compared to the three full hoods needed for the duration of the Soxhlet extraction. It is clear, both the microwave and pressurized fluid extraction methods use much less hood space for the extraction, making the extraction more energy efficient as far as hood usage than the Soxhlet extraction. More in depth energy usage and cost comparisons could be made but these are dependent on local utility rates. The Soxhlet extraction uses water for the condensers and electricity for the heating mantles during the 16–24 h extraction period. The microwave extraction uses electricity for the 20–45 min extraction time only. Pressurized fluid extraction uses electricity for heating the oven during the entire sample set processing time. Since the Soxhlet uses water in addition to electricity and the extraction time is longer, the energy usage would be highest for this method, followed by the pressurized fluid extraction, then the microwave extraction using the least amount of energy.

5.6 Extraction time

The overall time needed for the extraction impacts the resources needed in terms of lab equipment, personnel time and turnaround time of the samples. Ultimately all

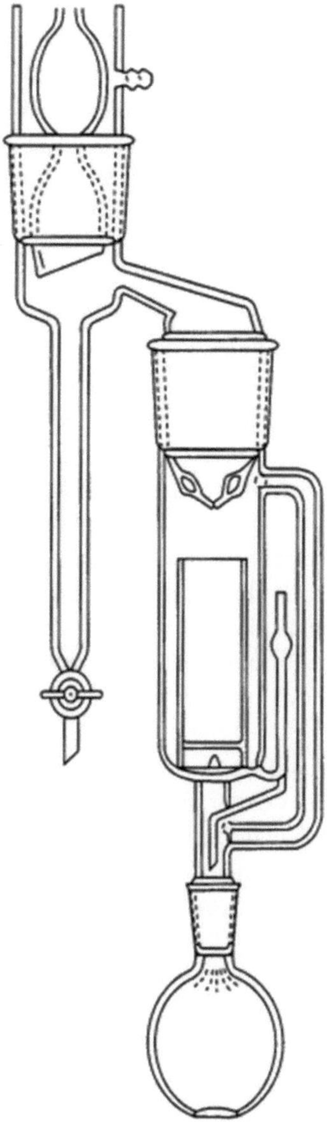

Figure 5.1: Soxhlet extractor with Dean Stark trap. Figure taken from EPA Method 1613b [1].

of these impact the cost of the analysis. Procedure includes the time needed to weigh the samples for extraction, which is the same for all methods, and the loading of the samples into the extraction apparatus is similar enough that it will be considered equivalent. Additionally, there are the concentration and cleanup steps that all of the extraction methods require and those are not considered as that time would be the same for all extracts regardless of the method employed for the extraction since the cleanup is a function of the sample matrix. However, with the pressurized fluid extraction method there is the potential to do some or all of the sample cleanup steps as the samples are eluted from the extraction vessels. It would depend on the sample matrix as to whether this would be enough to fully clean up the extract or if additional steps would be required. This aspect should be investigated for potential solvent, resource, time and energy savings. If the samples of interest are of similar matrix or enough is known about the sample matrices to be tested, this could be a key deciding factor on which method is optimal for a specific laboratory.

As stated previously, the Soxhlet extraction requires a minimum of 16 h up to 24 h for the actual sample extraction. Glassware setup and cleaning steps prior to loading the sample for extraction require approximately 4–5 h for a set of 20 samples. Once the glassware is assembled for the extraction, it must be loaded with clean solvent and the extraction is started to pre-clean the assembled glassware before any sample is loaded. This solvent is then discarded before the actual sample extraction is begun. At the start of the extraction and possibly periodically during it, an analyst must monitor the amount of water being collected in the Dean Stark trap and the water may need to be removed by draining the trap off periodically during the 16–24 h extraction. This manual intervention is a complication to an otherwise automated extraction that almost mandates a second or split shift operation in order to be able to monitor the sample extraction process properly.

There would also be significant glassware cleaning after the extraction before another set of samples could be started. This time would depend on whether the washing is handled manually or with an automated dishwasher. Some of the glassware used is fairly complex and delicate and therefore may have to be washed manually. This washing is in addition to the cleaning steps with solvent mentioned above. All of these steps also add time and cost to the process. A full set of 20 samples would be extracted over a 2 day period with another day before the glassware could be used again. However, depending on the personnel available the turnaround time could be even longer as that person may be performing the sample extract cleanup steps of the process and not be available to wash the glassware for another day or two depending on the steps necessary to clean up a particular sample matrix.

In contrast to this is the pressurized fluid extraction. The extraction vessel used in this case is a stainless steel cell with frits at both ends and stainless steel screw on caps. The extracts are collected in disposable glass vials. The extraction vessels will need to be cleaned after each extraction. Automatic dishwashing may be suitable in this case as none of the parts are particularly fragile with the possible exception of the frits. The extraction can be performed either during the day or overnight as the samples are processed sequentially and require approximately 5 h for the extraction of a set of 20 samples. The overnight extraction may be preferable from a time standpoint as the extracts can be ready for the cleanup at the start of the next day. The pressurized fluid extraction method takes less time to set up and to get started than the Soxhlet does. It would require only an hour or so before the extraction starts, so only a portion of the afternoon is needed to set up the samples for extraction with the extracts being immediately available the next morning to start the cleanup. It does take overnight, however, less analyst time is needed to start and monitor the extraction process. There is no requirement to monitor it overnight as the extraction equipment is fully automated and self-contained. Any water extracted is removed during the drying step.

The microwave extraction method uses disposable glass vessels thereby reducing the need for equipment washing in between uses. The microwave extraction method is capable of simultaneously extracting up to 24 samples. This significantly reduces the overall time needed for the extraction. The extraction time is about 20–30 min with a full cycle time including cool down which is approximately 45 min. That makes the microwave extraction the fastest method by far for a complete set of samples. The extracts can be prepared and the cleanup started in the same day. This reduces the cycle time by at least 1 full day, if not 2, over the other 2 extraction options. This would reduce the cost of analyst time, lower the turnaround time to the client and potentially require less lab equipment and resources than the other extraction methods. See Table 5.3 for a summary of the three extraction methods.

Table 5.3: Summary of extraction methods.

Soxhlet	PFE	ME
• Labor intensive • Very high solvent volume • Multiple hoods required for extensive time • Long extraction time • Low start-up cost of equipment • Long sample turnaround time	• Automated • Lower solvent volume • No hood required during extraction • Medium extraction time • High start-up cost • Medium sample turnaround time	• Automated • Lower solvent volume • No hood required, needs a permanant exhaust vent • Fast extraction time • High start-up cost • Fast sample turnaround time

5.7 Conclusion

After reviewing the solvent, hood usage, and time requirements used to compare the green impact of the various EPA approved extraction method choices available, the Soxhlet extraction method is clearly the least desirable choice. The other two methods have some advantages and disadvantages that may make one or the other more desirable depending on the situation in any given laboratory.

The Soxhlet extraction is the worst choice in terms of solvent consumption, hood usage requirements, analyst time and overall time. Its only advantages are that it has a relatively low start up cost of the equipment and that it has a long history of acceptable performance.

The pressurized fluid extraction uses less solvent, has no hood or ventilation requirements, uses disposable glassware, takes less time, and may be able to perform the sample cleanup steps as well as the extraction. Its disadvantages are the higher initial equipment cost and slightly longer overall extraction time.

The microwave extraction uses less solvent as well, uses disposable glassware, and takes the least amount of time for the extraction. The disadvantages of it are that it needs a permanent exhaust vent and also has higher initial equipment cost.

Whichever method is chosen, it needs to meet the performance criteria of EPA method 1613b in order to be acceptable for the purpose given. Based on the above factors in the NSF International laboratory, we would select the microwave extraction because of the lower solvent usage, the minimal hood requirements and the overall speed of the extraction. Giving considerations to factors related to Green Chemistry when selecting a method, as has been done here, allows a laboratory to make the most sustainable decision based on that particular laboratory's situation.

Acknowledgements: The author wishes to thank the following people for their assistance in researching the method and editing the chapter: Mr Daniel Colgan, Mr Kyle Barocko, Mr Timothy Baker and Dr Kerri LeVanseler.

References

[1] EPA method 1613 tetra- through octa-chlorinated dioxins and furans by isotope dilution HRGC/HRMS. Oct 1994. U.S. Environmental Protection Agency, Office of Water Engineering and Analysis Division (4303), 401 M Street SW Washington D.C. 20460.

[2] SW-846 - test methods for evaluating solid waste: physical/chemical methods, method 3545a. Available at: https://www.epa.gov/hw-sw846/sw-846-test-method-3545a-pressurized-fluid-extraction-pfe. Accessed: 22 Feb 2018.

[3] SW-846 - test methods for evaluating solid waste: physical/chemical methods, method 3546. Available at: https://www.epa.gov/hw-sw846/sw-846-test-method-3546-microwave-extraction. Accessed: 22 Feb 2018.

[4] SW-846 - test methods for evaluating solid waste: physical/chemical methods, method 3550C. Available at: https://www.epa.gov/hw-sw846/sw-846-test-method-3550c-ultrasonic-extraction. Accessed: 22 Feb 2018.

[5] SW-846 - test methods for evaluating solid waste: physical/chemical methods, method 8280B. Available at: https://www.epa.gov/hw-sw846/sw-846-test-method-8280b-polychlorinated-dibenzo-p-dioxins-pcdds-and-polychlorinated. Accessed: 22 Feb 2018.

[6] SW-846 - test methods for evaluating solid waste: physical/chemical methods, method 8290A. Available at: https://www.epa.gov/hw-sw846/sw-846-test-method-8290a-polychlori nated-dibenzodioxins-pcdds-and-polychlorinated. Accessed: 22 Feb 2018.
[7] International Agency for the Research on Cancer, World Health Organization. IARC monographs on the evaluation of carcinogenic risks to humans, IARC, 150 cours albert thomas. 69372 Lyon CEDEX 08, France. Available at: http://monographs.iarc.fr/ENG/Classification/latest_classif.php. Accessed: 23 May 2018.
[8] EPA toxocological review of toluene (CAS No. 108-88-3) in support of summary information on the integrated risk information system (IRIS). Sep 2005. U.S. Environmental Protection Agency, Washington D.C. Available at: https://www.epa.gov/sites/production/files/2014-03/documents/toluene_toxicology_review_0118tr_3v.pdf. Accessed: 23 May 2018.
[9] EPA toxocological review of hexane (CAS No. 110-54-3) in support of summary information on the integrated risk information system (IRIS). Nov 2005. U.S. Environmental Protection Agency, Washington D.C. Available at: https://nepis.epa.gov/Exe/ZyPDF.cgi/2000E79S.PDF?Dockey=2000E79S.PDF. Accessed: 23 May 2018.
[10] OSHA hazard communication, hazard classification guidance for manufacturers, importers and employers, occupational safety and administration. U.S. Department of Labor,OSHA 3844-02 2016. Available at: https://www.osha.gov/Publications/OSHA3844.pdf. Accessed: 23 May 2018.
[11] ThermoFisher scientific safety data sheet for methylene chloride, Revised 17 Jan 2018 Rev 6. Available at: https://www.fishersci.com/shop/msdsproxy?storeId=10652&productName=D371&productDescription=METHYLENE. Accessed: 23 May 2018.
[12] ThermoFisher scientific safety data sheet for toluene, Revised 17 Jan 2018 Rev 4. Available at: https://www.fishersci.com/shop/msdsproxy?productName=T326P4&productDescription=TOLUENE. Accessed: 23 May 2018.
[13] ThermoFisher scientific safety data sheet for hexane, Revised 19 Jan 2018 Rev 3. Available at: https://www.fishersci.com/shop/msdsproxy?productName=AC326922500&productDescription=N%2520HEXANE&catNo=AC32692-2500&vendorId=VN00032119&storeId=10652. Accessed: 23 May 2018.
[14] ThermoFisher scientific safety data sheet for acetone, Revised 19 Jan 2018 Rev 6. Available at: https://www.fishersci.com/shop/msdsproxy?productName=AC177170010&productDescription=ACETONE. Accessed: 23 May 2018.
[15] ThermoFisher scientific safety data sheet for methanol, Revised 18 Jan 2018 Rev 6. Available at: https://www.fishersci.com/msdsproxy%3FproductName%3DA412SK4%26productDescription%3DMETHANOL%2BACS%2B4L%2BSAFE-COTE%26catNo%3DA412-SK4%2B%26vendorId%3DVN00033897%26storeId%3D10652. Accessed: 23 May 2018.
[16] Cost comparison: budgeting for a filtered fume hood, Available at: http://www.labconco.com/articles/cost-comparison-filtered-fume-hood. Accessed: 25 May 2018.
[17] Gilly Q, Validating cost and energy savings from harvard's shut the sash program. Available at: https://green.harvard.edu/sites/green.harvard.edu/files/FumeHoodWhitePaper.pdf. Accessed: 25 May 2018.
[18] Validating cost and energy savings from Harvard's shut the sash program. Available at: http://sustainability.tufts.edu/wp-content/uploads/FumeHoodPaper.pdf. Hood reference not used at this time: http://sustainability.tufts.edu/wp-content/uploads/FumeHoodPaper.pdf. Accessed: 25 May 2018.

Joseph C. Furgal and Chamika U. Lenora

6 Green routes to silicon-based materials and their environmental implications

Abstract: The "greening" of silicon chemistry is fundamentally important for the future of the field. Traditional methods used to make silicon-based materials rely on carbon rich processes that are highly energy intensive, cause pollution, and are unsustainable. Researchers have taken up the challenge of developing new chemistries to circumvent the difficulties associated with traditional silicon material synthesis. Most of this work has been in the conversion of the "green" carbon neutral biogenic silica source rice hull ash (RHA, ~85 % silica) into useful silicon building blocks such as silica's, silicon, and alkoxysilanes by using the inherently higher surface area and reactivity of RHA to sidestep the low reactivity of mined silica sources. This is a review of the work that has been done in the area of developing more environmentally benign methods for the synthesis and use of silicon containing materials to eliminate the negative impact on the environment.

Keywords: biogenic silicon, rice hull ash, green silicones, silsesquioxanes, alkoxysilanes, silicone degradation

6.1 Introduction

Silicon-based materials have their origins in the Stone Age, when our ancestors used quartz and other silica-based stones to make tools (Innovations of Silicone 2015). It was not until ancient Rome that sand was used to make glass bowls and vases, beginning the modern era of silicon technologies. Silicon as an element was first isolated by J.J. von Berzelius in the early 1800s, and in the mid-1800s Sainte-Claire Deville developed the now common process of making silicon-based materials, carbothermal reduction, to synthesize pure silicon metal (Innovations of Silicone 2015).

During the 1890s, Frederick Kipping began studying the interactions of silicon and carbon, pioneered the development of organosilicon chemistry and was the first to coin the term silicone (Asimov, 1964; Klosowski & Wolf, 2009). In the early

This article has previously been published in the journal Physical Sciences Reviews. Please cite as: Furgal, J. C., Lenora, C. U. Green routes to silicon-based materials and their environmental implications *Physical Sciences Reviews* [Online] 2020, 5. DOI: 10.1515/psr-2019-0024.

https://doi.org/10.1515/9783110597783-006

twentieth century, Kipping and others were searching for a silicon-based derivative of carboxylic acids (siloxylic acid). However, their efforts were unsuccessful (not existent in any isolable quantity) and instead found a new class of materials that eventually became known as silsesquioxanes $[RSiO_{1.5}]_n$.

The biggest push for research in siloxane chemistries began in the 1930s when Andrianov hydrolyzed alkyltriethoxysilanes in water; followed by the hydrolysis of phenyltrichlorosilane to make the first isolated phenyl-silsesquioxane derivatives as a polymeric white solid resin (Zavin & Muzafarov, 2007; Andrianov, 1965).

The need for natural rubber alternatives during World War II pushed silicon chemistry much further, where such materials as silsesquioxane/silicone hybrid materials were found to offer ideal natural rubber replacements. One commercially successful discovery was the development of children's toy Silly Putty (Rosenberg, no date; Colas & Curtis, 2004).

Research on siloxanes continued full force through the 1940s and 50s through the heyday of polymer chemistry by the likes of Corning, General Electric, Dow-Corning, Wacker, etc. Today siloxane-based materials and their more basic counterparts, silica and silicon metal, are used in nearly every industry on the planet, from cosmetics to food, and from automotive to aerospace (Reisch 2011).

Though silicon-based materials are very important for many industries and are used in a diverse set of applications ranging from computer chips to coatings/sealants to adsorbents, the synthesis methods typically still rely on carbothermal reduction techniques developed in the mid-nineteenth century. These methods are extremely energy intensive and highly polluting compared to some newer methods discussed later in this paper, but continue to be used because the manufacturing processes are already in place and are expensive to replace. Even so, many process "greening" methods have been undertaken by researchers over the years and will be discussed in detail throughout this paper. Importantly, the use of cyclic siloxanes and their environmental impact have been more closely scrutinized, especially in the European Union where decisions have been made to ban the use and/or remaining residues of the cyclic siloxanes D_3, D_4, D_5 and D_6 in most consumer products. Many companies, especially in the cosmetics industry, are scrambling to find alternatives to the siloxanes that have become ubiquitous in the industry (Reisch 2011, 2012; Commission Regulation (EU) 2018/1516 2018; ECHA 2018).

The remainder of this paper will focus on (1) Conventional versus green chemical methods currently being utilized to produce silicon-based materials. (2) Conventional vs. environmentally friendly routes to alkoxysilane building blocks. (3) Siloxane polymers and copolymers synthesized and derivatized to be more earth friendly. (4) Degradation pathways of silicon-based materials under biological and environmental conditions.

6.2 Synthesis of silicon-based materials

6.2.1 Commercial methods to silicon-based materials

Alkoxysilanes and/or chlorosilanes are the major building blocks for nearly all silicon containing materials made today, including solar grade silicon, silicones, silsesquioxanes and various silicas, and found in applications ranging from toothpaste abrasives to photovoltaic devices for energy conversion (Wright and Uddin 2012). The most commonly used method to make functionalized silicon is to convert silica (i.e. mined sand) into metallurgical grade silicon (Si_{met}) through a carbothermal reduction process (Reaction 1).

$$SiO_2 + 2C \xrightarrow{1900°C} 2CO + Si_{met}(\text{met} = \text{metallurgical grade}, \approx 98\,\%\,\text{purity}) \tag{6.1}$$

$$Si_{met} + HCl \xrightarrow{\text{Cu/Sncatalyst/300°C}} HSiCl_3 + SiCl_4 \tag{6.2}$$

$$Si_{met} + EtOH(MeOH) \xrightarrow{\text{catalyst}} Si(OEt)_4\left[Si(OMe)_4\right] + H_2 \tag{6.3}$$

$$Si_{met} + MeCl \xrightarrow{\text{Cu/Sncatalyst/300°C}} MeSiCl_3 + Me_2SiCl_2(\text{silicone/polysiloxane precursor}) \tag{6.4}$$

$$4HSiCl_3 \xrightarrow{\text{disproportionaton catalyst}} SiH_4 + 3SiCl_4 \tag{6.5}$$

$$H_2 + HSiCl_3/SiCl_4(\text{or } SiH_4) \xrightarrow{\text{hot wire/rod}} Si_{pv}/Si_{eg}(99.99 + \%\text{purity}) + HCl(\text{recycled}) \tag{6.6}$$

$$SiCl_4 + 2H_2/O_2 \xrightarrow{>900°C} SiO_2(\text{fumed silica}) + 4HCl \tag{6.7}$$

This immensely expensive and energy intensive process requires heating silica to >1900 °C with carbon (coal) in an electric arc furnace, with the generation of CO and CO_2 as byproducts. Metallic silicon provides the basis for most silicon-containing materials, as outlined in reactions 1–7 (Gurav et al., 1993; Okamoto, 1994; Dosaj et al. 2000; Falcone 2007; Kawamoto and Okuwada 2007).

Such extreme conditions are necessary for the conversion of silica to Si_{met} because of the strength of the silicon-oxygen bond at 534 kJ/mol, one of the strongest naturally occurring bonds. Furthermore, silicon for photovoltaic applications must be extensively purified through the Siemens process, where Si_{met} is converted to $SiCl_4$ or SiH_4 and then re-reduced to Si_{PV} (Reactions 5 and 6), where HCl is luckily recycled. However, since chlorosilanes and $HCl_{(g)}$ are highly corrosive, and toxic, production plants require extensive safeguards, adding to already expensive economic and ecological costs.

In the synthesis of chloro-silanes, $HCl_{(g)}$ is combined with Si_{met} over a Cu/Sn catalyst (Reaction 2), to make both $HSiCl_3$ and $SiCl_4$. $SiCl_4$ could then be converted to fumed silica and Si_{pv} via Reactions 6–7. To make R-organic substitutions on the silicon (R_xSiCl_y) is a bit more complex. One method is to take an alkyl/aryl halide and create a reaction with H_xSiCl_y in a gas phase condensation reactor (Huang et al. 2013). An alternative method is that shown in Reaction 4, where Si_{met} is reacted directly with alkyl/aryl halides over a Cu/Sn catalyst at 300+ °C. The resulting products are R_xSiCl_y and HCl, with the different R_xSiCl_y derivatives being separated by distillation.

Functional materials can be made directly from the R-chlorosilane derivatives (i. e. D-type silicones (from R_2SiCl_2), T-type silsesquioxanes (from $RSiCl_3$), and M-type silanes R_3SiCl). Alternatively, the R-chlorosilanes can be functionalized further to their alkoxysilane (with alcohols) or cyclic siloxane derivatives for chlorine free processing. This process is achieved by refluxing R-chlorosilanes in methanol/ethanol with a RO^-Na^+ base catalyst, which through displacement of the chloride ion results in R-alkoxysilanes (Bonitz & Pfalz, 1970; Laine & Roll, 2011). R-alkoxysilanes are then used in applications such as sol-gel synthesis, silicones, and silsesquioxanes (Abe and Gunji 2004). *R-alkoxysilanes are more desirable than their R-chlorosilane precursors because of their increased stability to hydrolysis, decreased toxicity, and decreased corrosive properties* (Yanagisawa et al. 1989; Arkles et al. 1992; Jean et al. 2006).

Reaction 3 offers an alternative route to alkoxysilanes, where Si_{met} is combined with ethanol/methanol over a catalyst to make tetra-alkoxysilanes. These may be further functionalized through organo-Grignard/Lithium reagents or hydrolyzed directly make materials such as fumed/precipitated silicas (Yi et al. no date; Falcone 2007).

Overall, the energy intensive and environmentally unfriendly processes involved in deriving useful silicon materials are counterintuitive. The process is nearly cyclic in that it starts with oxidized silicon in the form of SiO_2 (sand), which is then converted to Si_{met}, this is then often converted back to fully oxidized materials such as fumed silica and alkoxysilanes. Due to this long and seemingly illogical process, many researchers have looked at methods to “shortcut” this process and make useful silicon-based materials directly from SiO_2 and its derivatives. The most promising method of doing so relies on biogenic silica from sources such as rice hull ash (RHA).

6.2.2 Silicon-based materials from bio-derived sources: rice hull(husk) ash (RHA)

One of the most promising areas of silicon chemistry research, enabling a decreased environmental impact, is the use of the raw material RHA. Over 134 million tons of RHA is produced per year, making it highly available, as well as being a ~90% amorphous silica (Quispe et al. 2017). Rice hulls are a major bi-product of rice production (670 million tons/year), and their husks can contain up to 20 % silica deposited through a biosilification process, by which plants such as rice absorb silica from the soil and store it in their stalks for structural support (Chandrasekhar et al., 2003; Asuncion et al., 2005; Shen et al. 2014). Kinrade et al. (1999) have suggested this takes place through pentacoordinated silicates imported and transported through the plant. It is not an entirely green process to make RHA, since the rice husks must be burned first or transformed through gasification in a biomass-energy conversion procedure, which releases CO_2. It is important to note that this industrial process to generate electricity is superior to the typical disposal method, which is to put the rice husks in a pile and burn them with an open flame, releasing small silica particles into the atmosphere, or by filling up landfills (Rodrigues, 2003; Rozainee et al., 2008b). It has been calculated that using rice hulls for energy generation would result in a near net zero CO_2 emission due to its derivation from bio-sources (Marchal et al. 2015). For more information on rice hull energy processes please see the review by Quispe et al. (2017) as well as work by Della et al. (2002), Kapur (1985), Hamad and Khattab (1981) and Rozainee et al. (2008a, 2008b). In this process the resulting energy can be used for other purposes vs. the carbothermal reduction methods, and less energy is then needed to convert the resulting RHA material to useful silicon-based materials since it's higher surface area increases reactivity (Della et al. 2002; Bakar et al. 2016). There are many companies throughout the world today converting rice hulls to energy with RHA being stockpiled (Tullo 2015). RHA can be used as a source material in place of the more processed silica starting materials mentioned above that take inordinate amounts of energy to convert. Additionally, RHA may also be used directly as a silica-blending agent, without the need for conversion, as well as production of industrial cements (Rodrigues 2003), tires, (GOODYEAR REACHES SUPPLY AGREEMENTS FOR RICE HUSK ASH SILICA 2015) batteries (Liu et al., 2013; Wong et al., 2014; Pode, 2016), catalysts (Adam et al. 2012), insulation (Kalapathy et al. 2003; Pode 2016), and water purification (Adams and Mulaba-Bafubiandi 2014).

6.2.3 RHA as a silica source

One of the major challenges in using RHA as a silica source is the effectiveness of residual carbon removal strategies. In typical processes, silica is either extracted with superheated steam after a pyrolysis step to remove residual carbon by generating hydrogen, carbon dioxide and monoxide or by a method in which an acid such as HF is used in conjunction with ammonia to give ammonium fluoride and silica of high purity (99.9 %) after rinsing (Todkar et al. 2016). Many groups over the years have focused on improving this process, though most are still only on the cusp of green chemistry. For example, many groups have sought to use base derived techniques instead of acid due to the easier conversion and extractability of the resulting silicates or silicas. For example, An et al. (2010) have developed a strategy to extract silica from RHA (acid pretreated) using sodium carbonate as a base followed by carbonation at 90 °C to remove carbon dioxide from the mixture and generate pure silica. The procedure can also be reversed, in which sodium hydroxide is added first to an RHA emulsion in water to form sodium silicates followed by precipitation with sulfuric acid and washing (remove sulfates) to give silica (Todkar et al. 2016). An alternative method to biogenic silica recently developed by Schneider and co-workers uses a two-step procedure in which the rice husk (or other biosilification plant material) is subjected first to a solution of citric acid and water under vigorous stirring as a pretreatment, followed by a sequential and stepwise thermal heating cycle up to 550 °C over 5.5 hours (Schneider et al. 2018). They were able to obtain silica with up to 99.8 % purity from this process, and surface areas up to 300 m^2/g, which is comparable to conventional approaches. Though their process heats to a moderately high temperature, their pretreatment steps using a bio-synthesized acid and no toxic chemicals gives high credence to being a green chemical method. Table 6.1 compares some common methods for the generation of silica from biogenic sources (Schneider et al. 2018).

Table 6.1: Literature overview of leaching-burning procedures of selected biomasses to obtain biogenic silica (Schneider et al. 2018).

Biomass	Leaching agent	Leaching conditions	Purity ash	S_{BET} (m^2 g^{-1})
Rice husk	Distilled water	50 °C/1 h	97.3 wt%	n.d.
	1–7 wt% citric acid	25–80 °C/1 h	99.5–99.8 wt%	n.d.
Rice husk	3 % (v/v) HCl	100 °C/2 h	99.2 wt%	321
	10 % (v/v) H_2SO_4	100 °C/2 h	99.6 wt%	282
	Distilled water	100 °C/2 h	95.5 wt%	194
	3 % (v/v) NaOH	25 °C/24 h	39.8 wt%	n.d.
Rice husk	0.01–1 N HCl	100 °C/1.5 h	95.3–95.7 wt%	221–260
	0.01–1 N HNO_3		94.1–97.8 wt%	217
	0.1–5 N acetic acid		95.0–96.0 wt%	193–215
	0.1 N-sat. oxalic acid		94.8–95.5 wt%	167–271
Rice husk	10 % (v/v) acetic acid	150 °C/1 h	98.7 at.%	n. d.
	10 % (v/v) citric acid		98.8 at.%	n. d.
	0.2 M H_3PO_4		98.8 at.%	n. d.
Oat husk	0.1 N HCl	90 °C/1 h	94.1–94.3 wt%	124–129
Rice straw	0.1 N HCl	90 °C/1 h	93.4–97.5 wt%	310
Horsetail	10 % HCl	100 °C/2 h	n. d.	233
Horsetail	2 % H_2SO_4	100 °C/1 h	98.3 wt%	393

Sat. Saturated *n.d.* not determined.

6.2.4 RHA as a source for silicon and silicon carbide

Silicon metal and silicon carbide are two extremely useful materials as both function as semi-conducting materials. Both are typically synthesized by conventional carbothermal reduction processes discussed above; however, it is possible to also make these electronically important materials using RHA.

One of the earliest, albeit not so environmentally friendly, instances of the conversion of RHA to silicon metal comes from work by Hunt et al. in which they coked acid leached rice hulls from the Quaker Oats Company at 800 °C for ~45 min. The total mass was then reduced by 35 % of the initial mass, with their impurities such as Al, K, and Mg analyzed (Hunt et al. 1982; Ikram and Akhter 1988). The subsequent ash was pelletized with sucrose to control the material density for conversion to silicon metal in an electric arc furnace. The coked RHA was found to have produced silicon with 1–2 orders of magnitude lower impurities than conventional approaches. This is due to the lower propensity for rice plants to take up heavy metals from the environment.

More recently, Okutani (2009) and Laine et al. (2013; Marchal et al. 2015) have devised better approaches toward solar/chip grade silicon. Okutani (2009) has developed an approach that converts RHA silica in a single step to $SiCl_4$ using carbon and chlorine gas and also a method to use RHA silica as a reactant with aluminum as a reducing agent to form Si_{met} and Al_2O_3 directly at 1300 °C over 20 min. Though the first step of this reaction is similar to the carbothermal reduction process discussed above, the higher surface area of the RHA allows for a lower temperature and more efficient reaction by pyrolyzing the carbon present in RHA at 1000 °C under argon in an electric arc furnace followed by a switch to chlorine gas for chlorination. This method resulted in 80 % conversion of RHA silica to $SiCl_4$ in 1 hour, whereas typical amorphous silica/activated carbon systems under the same conditions resulted in only 3 % conversion. The process of converting RHA silica directly to Si_{met} using an aluminum reducing agent is less attractive, but still works reasonably well, resulting in >70 % conversion of silica to Si_{met} in around 30 min, though the purity suffered.

Laine et al. offer a greener less energy intensive approach to up to 99.9999 % pure silicon (Marchal et al. 2015). The first step was to mill RHA in a 3.7 wt% HCl aqueous solution heated to 200 °C and then hot water washes to remove persistent impurities such as phosphorous and alkali metals that lead to lower purity silicon. They then used a 50 kW electric arc furnace in a batch process with acid washed RHA silica pelletized with graphite to generate Si_{met}. The process can convert 3.6 kg of pellets to 1.6 g of silicon every 6 hours. This process could greatly reduce the environmental impact of synthesizing silicon for energy capture, storage and photonics, which are currently made by standard carbothermal reduction processes.

SiC can also be formed directly from RHA as pioneered in the mid-1970s by Lee and Cutler (Lee and Cutler 1975; Lee et al. 1977). Krishnarao et al. (Krishnarao 1992a, Krishnarao 1992b; Krishnarao et al. 1998) as well as others (Mizuki et al. 1993; Sujirote and Leangsuwan 2003; Niyomwas 2009; Li et al. 2013) have developed methods for the direct conversion of RHA to SiC using less energy intensive processes. This is typically done by first pretreating the RHA with acid to remove impurities, via direct RHA insertion, or by pre-coking the samples. This is typically followed by pyrolysis to reduce the carbon-silicon mixture into a ceramic under an inert atmosphere in a furnace pressure chamber at ~1425 °C, since unlike with Si_{met} the carbon should be trapped. Reactions 8–10 outline the general chemical approach. This method usually

$$SiO_2 + C \rightarrow SiO + C \tag{6.8}$$

$$SiO + 2C \rightarrow SiC + CO \tag{6.9}$$

$$SiO_2 + 3C \rightarrow SiC + 2CO \tag{6.10}$$

results in SiC whiskers, which are useful for adding strength to composite materials. α-SiC and β-SiC are also possible as well as silicon nitrides (Krishnarao et al. 1998; Sujirote and Leangsuwan 2003). As with the formation of silicon metal, the processes associated with forming SiC are not particularly environmentally friendly, but should be overall close to carbon neutral when counting all energy in and energy out. The purification of RHA is estimated at 5 kWh/kg (vs. final silicon produced), whereas the energy from rice hulls can reach 23 kWh/kg vs. final silicon produced (Marchal et al. 2015).

6.2.5 Applications of RHA derived materials

RHA has found use in many different applications, the most prevalent being for the generation of various forms of amorphous silica which find uses as ingredients in food, pesticides, personal care products by use as structural fillers, adsorbents and active abrasives in polymers and ceramics (Sun & Gong, 2001; Pode, 2016). This area has also been the subject of most papers on RHA and contains a comprehensive review by Sun and Gong (Kapur, 1985; Kamath & Proctor, 1998; Proctor et al. 2000; Della et al. 2002; Liou 2004; Tang and Wang 2005; An et al. 2010; Wang et al. 2011; Ma et al. 2012; Shen et al. 2014; Bakar et al. 2016; Schneider et al. 2018). Amorphous silicas are highly sought after due to their higher surface area and thus reactivity over the much more earth crust abundant crystalline silica (Asuncion et al., 2005; Wang et al., 2011). The techniques for generating silica from RHA in most cases mirror those discussed in 2.2. In general, RHA is first acid washed to remove impurities and then pyrolyzed at temperatures above 600 °C for up to 8 h. Amorphous silica is a major use in the production of tires, as it is added to reduce road wear by decreasing rolling friction in tires. Goodyear has now introduced tires with silica derived from RHA (Rubber and Plastics News 2015; GOODYEAR REACHES SUPPLY AGREEMENTS FOR RICE HUSK ASH SILICA 2015).

Concrete additives are one of the major areas being explored for direct use of RHA (Rodrigues, 2003; Foletto et al., 2006; Ghofrani et al., 2015; Fapohunda et al. 2017; Zareei et al. 2017). Portland cement is one of the most consumed materials on the planet resulting in ~1 ton of concrete/person/year. Cement is made up of a complex mixture of silicates (~75 wt%), aluminates and other calcium derived complex anions and is typically made using an energy intensive process with high temperature (1450 °C) and solid-state reactions from mined materials (Rodrigues 2003). The use of RHA can allow the formation of the main ingredient of cement (β-Ca_2SiO_4) at temperatures as low as 700 °C. By adding CaO to the RHA mixture during heating, calcium silicates are formed instead of silica.

RHA can also be used directly as an additive replacement for Portland cement (Fapohunda et al. 2017; Zareei et al. 2017). When RHA is added up to 15 % by weight of the mixture as a replacement for cement, it offered a better compressive load strength for the full concrete mixture vs. a micro-structured silica control of 56 MPa and 50 MPa, respectively (Zareei et al. 2017). An increase in tensile strength and also a decrease in chloride permeability was observed for the RHA concrete vs. the control.

RHA is being explored as a thermal insulation material (Kalapathy et al. 2003; Pode 2016). Kalapathy et al. used 1 M NaOH to turn RHA into silicates, which upon filtering to remove carbon could be boiled down to disk shaped blocks at three heating times (4, 6 and 8 h) to remove water. Table 6.2 shows the density; break force and conductivity of the materials. The overall density of ~0.33 is much lower than the 1.1 g/mol typically found with commercial silica xerogels, while the thermal conductivity (an important property for insulation) is slightly lower than that for styrene based insulation (0.103 and 0.12 $Wm^{-1}K^{-1}$, respectively).

Table 6.2: Density mechanical strength and thermal conductivity of silicate blocks produced by evaporation of water at three different heating rates.

Sample	Density (g cm^{-3})	Peak force at break (N cm^{-2})	Conductivity (W m^{-1} K^{-1})	Heating rate
Block 1	0.33 (0.015)	145 (3.6)	0.103 (0.002)	High
Block 2	0.38 (0.015)	161 (3.0)	0.113 (0.005)	Intermediate
Block 3	0.42 (0.025)	196 (8.6)	0.128 (0.003)	Low

Notes: Values in the parentheses are standard deviations of the corresponding measurements.

As environmentally friendly uses of RHA go, few garner as much hype as uses in batteries, which offers a double bonus (Liu et al., 2013; Wong et al., 2014). For example, Cui and coworkers have developed a process to make silicon nanoparticles (10–40 nm) from RHA for use in Li-ion battery anodes which offer seven times higher reversible capacity (2,700 mAh/g) 7x greater than graphite only anodes (Liu et al. 2013). The procedure is outlined in Figure 6.1, where the initial steps are very similar to typical RHA processing methods, followed by electrolytic reduction of nano-SiO_2 with magnesium to form porous nano-Si. In general the temperatures needed for these processes are much lower than traditional Si_{met}, due to the higher reactivity of RHA over non-biogenic silicas (sands). The porosity and small size of this nano-Si gives it a distinct advantage over standard bulk silicon by giving it far more reactive sites per unit area. This led to highly repeatable charge-discharge cycling with average capacities of 1750 mAh/g over 150 cycles with highly reversible Li insertion and extraction. By comparison, standard Si_{met} showed an average capacity of less than 500 mAh/g under the same cycling conditions.

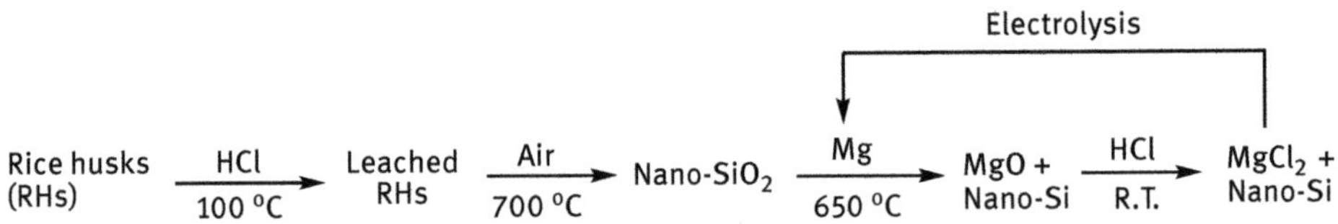

Figure 6.1: Synthesis of nano-silicon from RHA using oxidative pyrolysis and magnesium reduction (Liu et al. 2013).

Another application area being explored for RHA is its use as a purification material for water, an additional two for one green chemical approach. Adams and Mulaba-Bafubiandi have explored the use of RHA as an adsorbent for removing turbidity from water (Adams and Mulaba-Bafubiandi 2014). They explored two different methods of rice husk preparation (parboiled, partially cooked in the husk) and not parboiled; each was then rinsed, dried and incinerated in air in a porcelain crucible, followed by ashing at 800 °C. They analyzed the effective clarification by each method with a water sample with 88 NTU turbitidy, 6.63 pH and total dissolved solids of 127.3 ppm. They found that the higher surface area (but lower pore volume/sizes) of the parboiled RHA gave nearly a two-fold better reduction in dissolved solids but similar turbidity removal at 8 g loadings. All systems gave an increase in pH at all loadings.

Finally, RHA is being explored as a catalyst/catalyst scaffold for many reactions, and we direct the reader to the review article by Adam et al. to learn more (Adam et al. 2012).

As should be evident from the above discussion, RHA/other biogenic plant based green materials have become a very important silicon source for highly pure functional silicon-based materials. The development of many areas of silicon chemistry and engineering will continue to grow out of RHA sources as the extraction processes are improved and moved into main stream synthesis from a niche research area. With the positive publicity RHA has received from corporations such as Goodyear, it will become a more viable and visible silicon source as the high-energy requirements of processing silicon become vulnerable to planetary carbon reduction strategies.

In addition to the extensive base materials and many applications derived from RHA materials, a more detailed discussion of the synthesis of conventional alkoxysilane building blocks is also necessary. Section 6.2 discusses the beginnings of this methodology.

6.3 Beyond carbothermal reduction: other routes to alkoxysilane building blocks

6.3.1 Alkoxysilanes directly from silica

Researchers, as early as the 1930s, sought to develop greener methods to develop silicon building blocks (alkoxysilanes), especially through the direct low-temperature depolymerization of silica. Figure 6.2 illustrates the basic idea of this principle, in which a catalyst is used to lower the activation energy needed to break open the Si-O bonds in silica, followed by an exchange reaction with alcohol resulting in an alkoxysilane functionalized material, which depending on alkoxy group choice would result in distillable materials for easy purification.

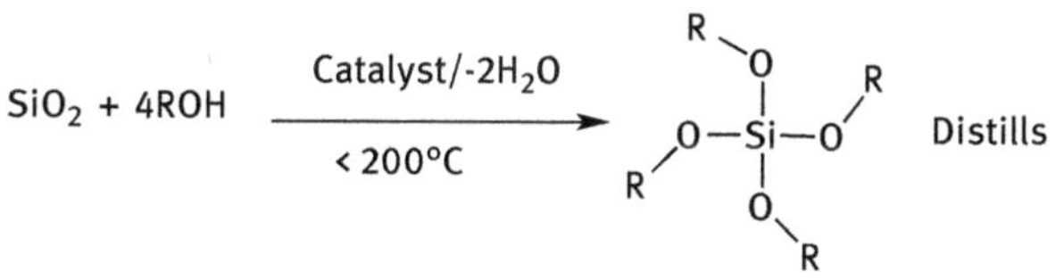

Figure 6.2: Depolymerization of silica by alcohol.

Rosenheim et al. in 1931 were the first to describe the process of silica depolymerization, in which they used catechol and stoichiometric KOH to form hexacoordinated triscatecholato silicates (Figure 6.3) (Rosenheim et al. 1931).

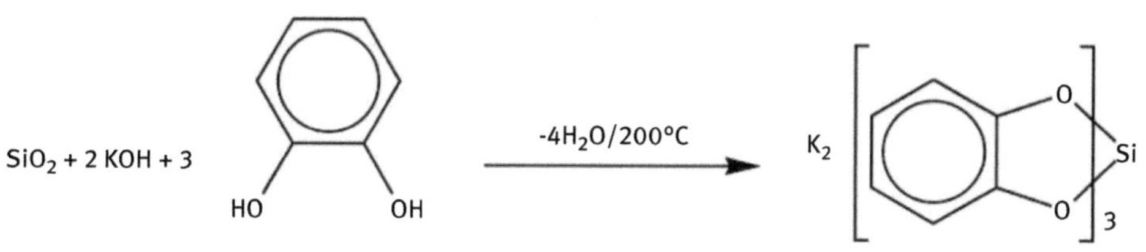

Figure 6.3: Depolymerization of silica with catechol and KOH (Rosenheim et al. 1931).

Though this material was not distillable and not particularly useful for many applications its importance was evidenced by many researchers, including Weiss et al. (1961),

Frye (1964), Boer et al. 1968, Flynn and Boer 1969, Barnum (1970, 1972) and Boudin et al. (1986), which continued to explore silica depolymerization, using Rosenheim's methods, but with various silica feedstocks.

The above reaction (Figure 6.3) is successful because, unlike carbon, silicon has the ability to expand its coordination sphere to accommodate five and six bonds, thus enabling decreased Si-O bond strength from electronic repulsions. Even though this process works quite well for depolymerizing silica at lower temperatures, the resulting hexacoordinated triscatecholato silicate was incredibly stable and also water-soluble. Since the reaction relied on a stoichiometric amount of base and diol (catechol), the use of it as an intermediate for the transformation of silica to other silicon products was unattractive.

In 1991 Laine et al. described an alternate route to diolato-silicates (Laine et al. 1991). This method used ethylene glycol instead of catechol to conduct the silica depolymerization process (Figure 6.4 and Figure 6.5). More recently, Laine et al. demonstrated that this process could be driven catalytically and is first order in base and surface area. The activation energy of this process is ~ 60 kJ/mol and works better with high surface area (>25 m^2/g), amorphous silicas like fumed silica or processed silica gels (Figure 6.6) (Blohowiak et al. 1994; Cheng et al. 2000). A major downside to this method, besides the use of toxic ethylene glycol, is that the product could not be distilled for easy purification and future use was limited to silica formation.

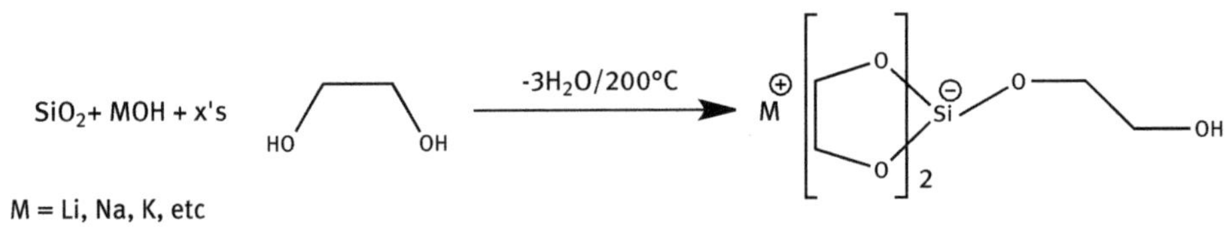

Figure 6.4: Depolymerization of silica with ethylene glycol and metal hydroxides to form glycolato-silicate (Si^{-1}) (Laine et al. 1991).

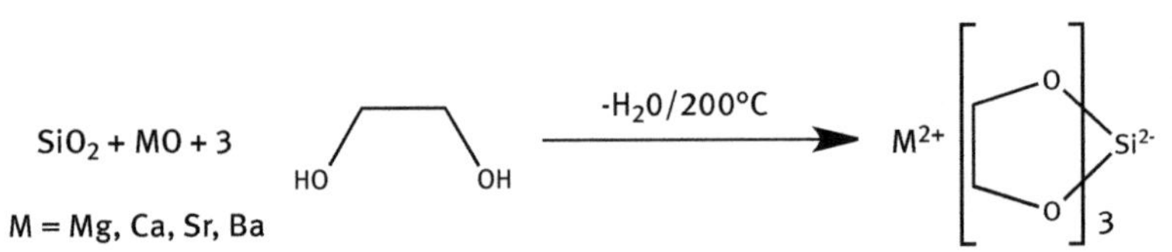

Figure 6.5: Depolymerization of silica with ethylene glycol and metal oxides to form glycolatosilicate (Si^{-2}) (Blohowiak et al. 1994; Cheng et al. 2000).

To improve upon the catalyzed ethylene glycol synthetic methods, other diols have now been used to develop distillable spirosiloxanes. It was found that sterically

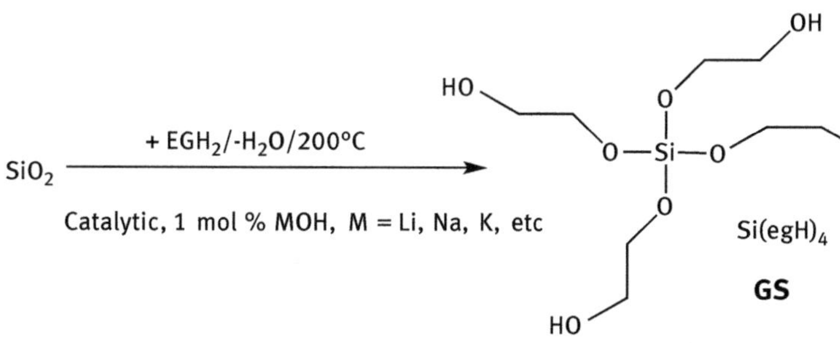

Figure 6.6: Depolymerization of silica with ethylene glycol to form glycolatosilicate.

hindered diols showed even better conversion and catalytic activity to spirosiloxanes (tetra-coordinated silanes) (Frye 1969). It was found that among those, 2-methyl-2,4-pentanediol, the hydrogenated product of base-catalyzed acetone condensation, which is also non-bio-accumulative and biodegradable, showed excellent conversions to Si(2-methyl-2,4-pentanediolato) $_2$ in 98 % isolated yield from fumed silica (350 m^2/g) and >60 % conversion of high surface area RHA (see below) to starting materials in less than 24 h (Figure 6.7) (Laine et al. 2016a, 2016b). This product is fractionally distilled from the reaction pot and separated from residual diol performing a hexane/water extraction.

Though efficient methods have been achieved to depolymerize silica, working with many of the spirosiloxane type derivatives can be difficult due to their lower reactivity and higher boiling by-products from hydrolysis-condensation methods used in Sol-Gel processing (Laine et al. 2016b). Therefore, efforts have been undertaken to convert or by-pass the need for spirosiloxane derivatives. Furgal et al. developed a simple process to convert the Si (2-methyl-2,4-pentanediolato)$_2$ derivative shown in Figure 6.7 to tetraethoxysilane (TEOS) or tetramethoxysilane (TMOS) by using trifluoroacetic acid catalyzed exchange of the diol with large excess of ethanol/methanol under anhydrous conditions (Figure 6.8) (Laine et al. 2016b).

Fukaya et al. have also developed a process to synthesize tetraethoxysilane directly, circumventing the need to go through the spirosiloxane intermediate (Fukaya, Choi,

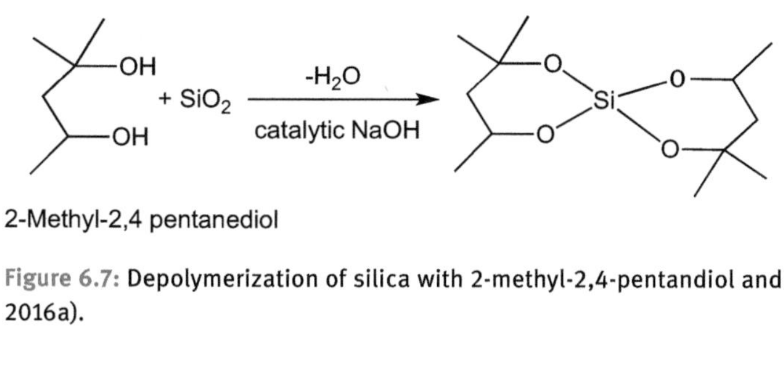

Figure 6.7: Depolymerization of silica with 2-methyl-2,4-pentandiol and catalytic NaOH (Laine et al. 2016a).

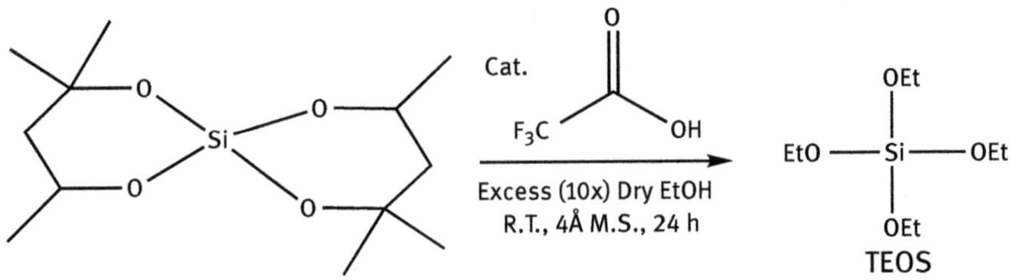

Figure 6.8: Synthesis of tetraethoxysilane (TEOS) from Si (2-methyl-2,4-pentanediolato)$_2$ (Laine et al. 2016b).

Horikoshi, Kataoka, Endo, Kumai, Hasegawa, and Sato, 2017). This method relies on the use of a pressure vessel containing molecular sieves in the head-space and alcohol, KOH and silica source in the pot (Figure 6.9 and Figure 6.10). Their method using 3 Å molecular sieves, ethanol, 10 mol% KOH and high surface area production silica at 260 °C led to conversions as high as 70 % TEOS in 6 h, and with natural high surface area RHA (224 m^2/g) giving nearly 60 % yield, with most of the remainder being di-siloxane (Figure 6.10). Using higher boiling alcohols such as *n*-PrOH and *n*-BuOH resulted in slightly higher yields (73 and 79 %, respectively) and a lower propensity toward di-siloxane formation.

Other methods to synthesize tetraethoxysilane directly include multiple works by Goodwin et al. that used the reaction between calcium silicates (anhydrous) and HCl/

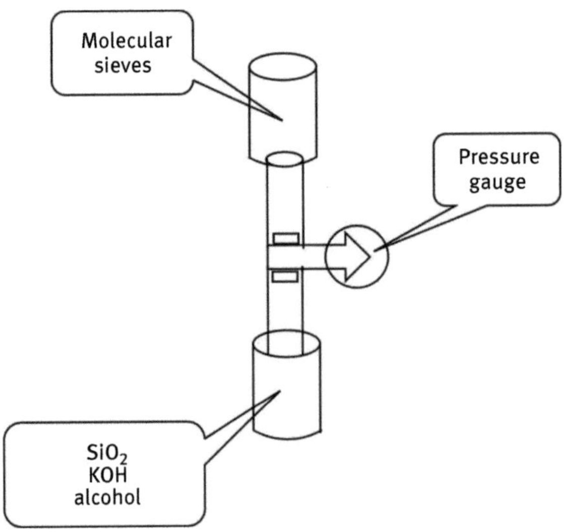

Figure 6.9: Reaction apparatus for the conversion of silica to TEOS in one-pot reaction (Fukaya, Choi, Horikoshi, Kataoka, Endo, Kumai, Hasegawa, and Sato, 2017). Redrawn and adapted from publication.

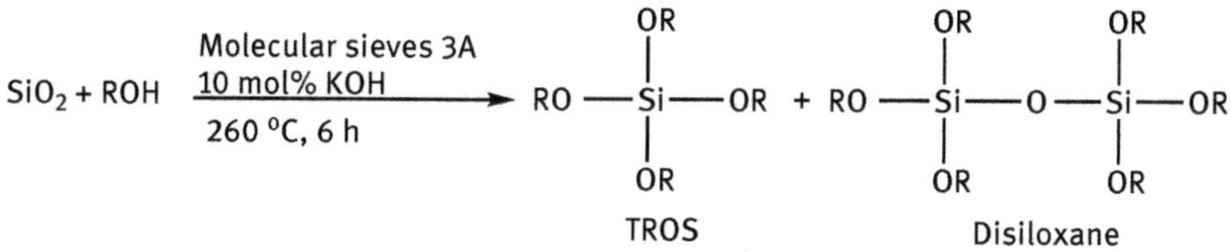

Figure 6.10: Catalytic reaction of silica with alcohol using the reaction apparatus in Figure 6.9 (Fukaya, Choi, Horikoshi, Kataoka, Endo, Kumai, Hasegawa, Sato and J. C. Choi 2017).

ethanol mixtures resulting 42% isolated TEOS yield (Kenney & Goodwin, 1988; Goodwin & Kenney, 1990). This process was relatively straight forward, but required extensive high-temperature drying of the silicates before reaction and resulted in many cyclic and dimeric siloxane side products. Other work by Bailey and O'Conner used dehydration methods through the azeotropic distillation from benzene/ethanol mixtures to generate TEOS (Bailey and O'Connor 1959). These methods, especially the latter, would not be considered extremely green processes.

A few methods have also been developed over the years to synthesize tetramethoxysilane directly from silica sources; however, these methods have either relied on sacrificial reagents resulting in a non-catalytic process (Figure 6.11) (Fukaya et al. 2016), or catalytic methods needing much higher temperatures (~500 °C) (Akiyama et al. 1993), circumventing the green chemical nature of the processes. Tetramethoxysilane is inherently more difficult to obtain directly due to its high reactivity with moisture, forcing the use of reactive organic drying agents such as 2,2′-dimethoxypropane or dimethoxycarbonate. The reaction of dimethylcarbonate with silica and alkali bases has been explored using a DFT quantum chemical investigation to aid in improving the reaction. They find that CsOH is the best hydroxyl source for SiO_2 conversion whereas LiOH is the poorest due to the formation of a stable dimethylcarbonate intermediate (Butera et al. 2018).

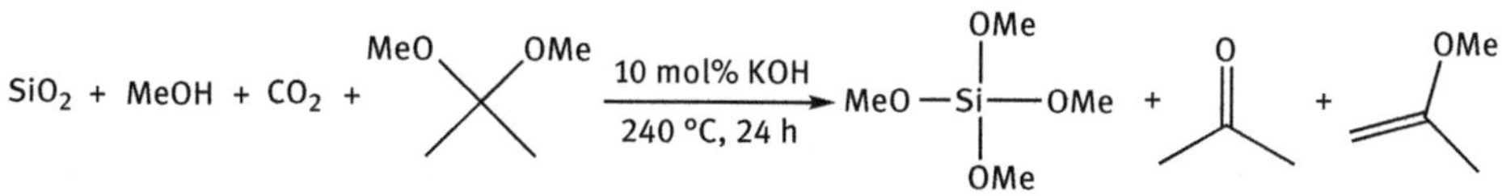

Figure 6.11: Conversion of silica to tetramethoxysilane pressurized CO_2 and a sacrificial organic drying agent (Fukaya et al. 2016).

Most recently, Sánchez-Ramírez et al. (2018), in two case study papers, have investigated and compared the environmental impact consequences, return on investment and purity/yield of TEOS synthesis/separation using a reactive distillation column (1-step) versus conventional reaction/separation procedures (2-step) processes (Figure 6.12). They find that though the reactive distillation column was more

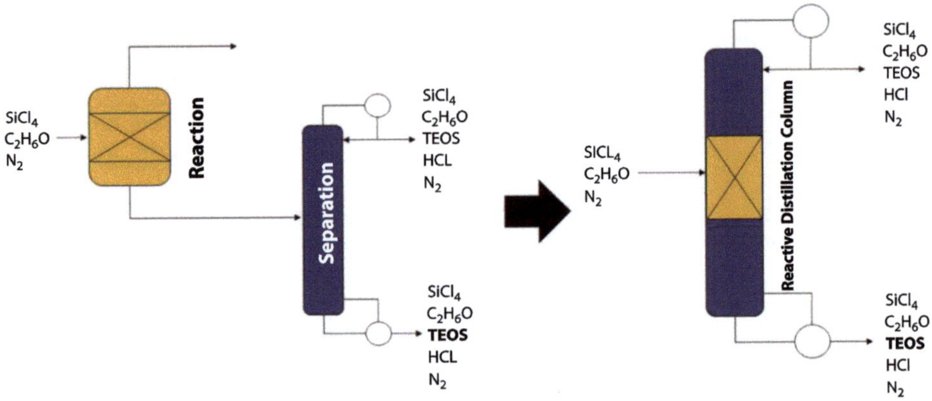

Figure 6.12: Conventional TEOS synthesis as derived from Reactions 1–7 above vs. new reactive distillation column methods (used with permission from the American Chemical Society) (Sánchez-Ramírez et al. 2018).

economical for many purities of TEOS, the 2-step reaction/separation procedure was more economical for the highest purity of TEOS (99.5 %+) and environmentally favorable due to the lower energy usage needed, reducing the green chemistry impact of this new process.

As is evident there are only a few methodologies that are relatively environmentally friendly for making alkoxysilane building blocks and the authors of this paper propose a call to improve these processes and methodologies for the future of silicon chemistry.

From these above methods a few groups have moved forward with developing methods to functionalize those basic building blocks for siloxane and silsesquioxane development by making R-alkoxysilane derivatives. These will be outlined in the next section.

6.3.2 R-alkoxysilanes from dialato silicates and spirosiloxanes

Efforts have been undertaken for the development of low-temperature routes to Si-C bonding systems that do not require the synthesis of Si_{met} as a starting material. A first step in this ideal direction is the establishment of new synthetic routes to Si-C containing compounds.

Surprisingly, few researchers have worked to tackle this goal. Most development in this area was carried out by Corriu et al, who explored the potential to react alkoxysilanes (hexacoordinated triscatecholato silicates) with nucleophiles such as Lithium and Grignard reagents to form Si-C bonds directly as illustrated in Figure 6.13

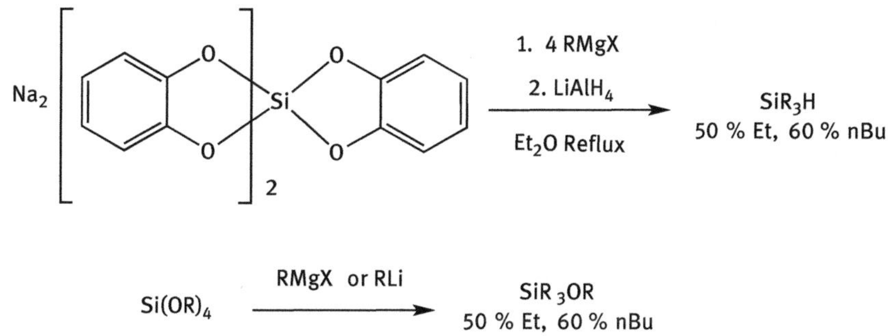

Figure 6.13: Typical reactions of triscatecholato silicate or tetralkoxysilanes with Grignard and lithium reagents (George et al. 1955; McBee et al. 1955; Boudin et al. 1986; Tour et al. 1992; Silverman and Rakita 1996; Klokov 2001; Jung et al. 2005).

(George et al. 1955; McBee et al. 1955; Boudin et al. 1986; Tour et al. 1992; Silverman and Rakita 1996; Klokov 2001; Jung et al. 2005).

Corriu et al. reported reasonable yields for these reactions; however, the only products are tri- and tetrasubstituted silanes, as opposed to the much more useful mono- or di-substituted derivatives. The latter compounds offer potential as monomer units for the synthesis of novel siloxanes and silsesquioxanes. The Corriu group concluded that the pentacoordinated/hexacoordinated intermediates formed are much more reactive than the tetracoordinated species, explaining why tri- and tetra-substituted silanes are the major products from their efforts versus mono- and di-functionalized compounds (Frye 1969; Boudin et al. 1986, 1988b, and 1988a).

DeShong et al. have been successful in generating mono-substituted R-groups at ethoxysilane centers using highly controlled low-temperature organolithium (R-Phenyl-Li) and Grignard (R-Phenyl-MgBr) reagents. These are used to nucleophilically displace an ethoxy group on TEOS (Figure 6.14) in good yield and high control

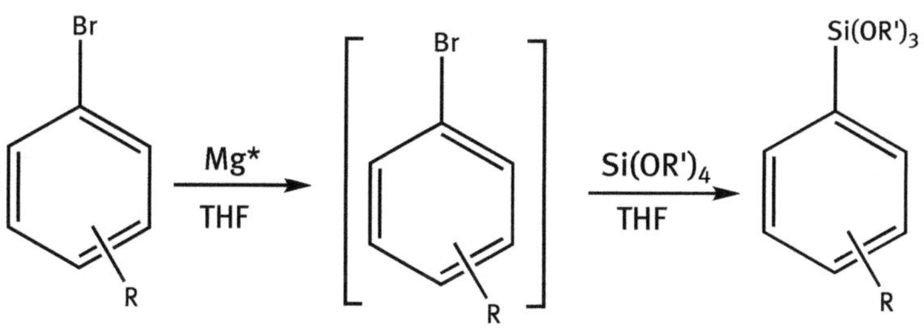

Figure 6.14: Synthesis of R-aryltriethoxysilanes by Grignard reaction followed by substitution at TEOS (Manoso et al. 2004). Redrawn from Manoso et al. 2004.

toward mono-functionalization at −78 °C in di-ethylether and tetrahydrofuran. Though their industrial TEOS source and organometallic reagent conditions do not fall under green chemistry standards, replacing the source materials and reaction conditions could improve the environmental impact of this reaction.

In order to take one step toward lessening environmental impact, Furgal and Laine synthesized a series of R-trialkoxysilanes derived from the Si(2-methyl-2,4-pentanediolato)$_2$ spirosiloxane mentioned previously (Figure 6.7). Their interest was inspired by the need to develop greener/more efficient routes to highly specialized and unique silsesquioxanes (Jung & Laine, 2011; Laine & Roll, 2011; Jung et al., 2012; Furgal et al. 2013, 2016); however, the ability to make siloxanes was also important. Their method, as outlined in Figure 6.15 (for Ph derivatives). used stoichiometrically controlled reaction conditions with a large array of R groups (R = Ph, thiophene, anthracene, 9,9-dimethylfluorene, etc.). Both R_1 and R_2 species were synthesized, with R_1 compounds requiring quenching of the intermediate free hydroxyl group with either MeI or Me_3SiCl before warming to room temperature and extraction (Furgal and Laine 2016).

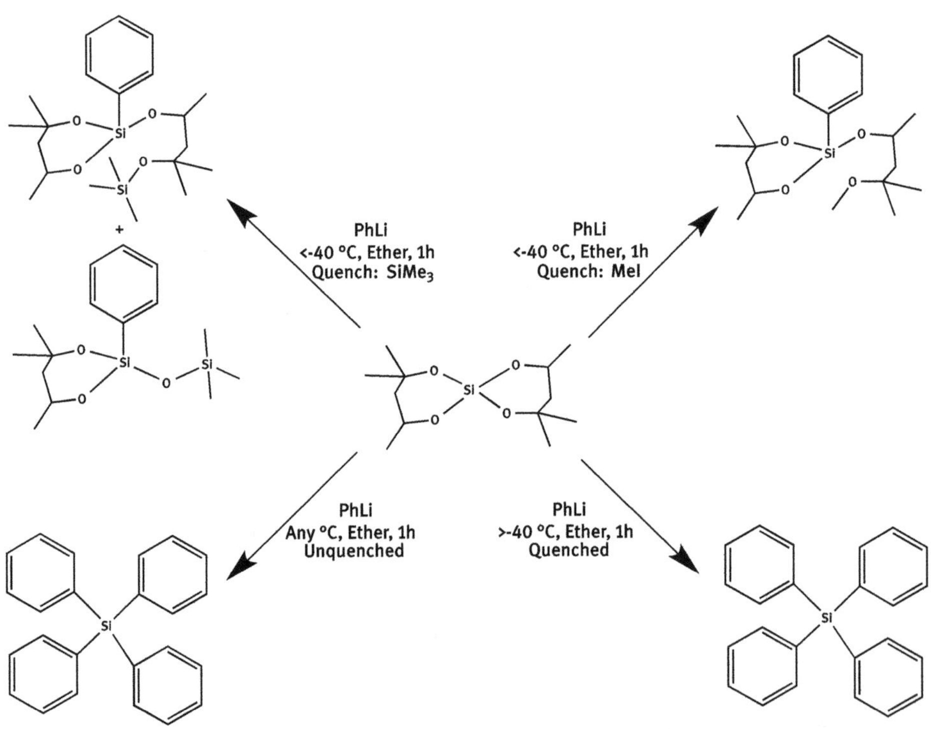

Figure 6.15: Example reaction conditions for generating different substitution products from Si(2-methyl-2,4-pentanediolato)$_2$ (Furgal and Laine 2016).

An obvious pitfall of all of the above Si–C bond forming reactions is the need to synthesize and use organo-Grignard and Lithium reagents, which are most commonly made using non-green solvent and reaction conditions (Hazimeh et al. 2006). There has, however, been advances in these methods in recent years which have mainly focused on the use of 2-methyltetrahydrofuran (Me-THF) as a more environmentally conscience alternative solvent to common solvents such as tetrahydrofuran and di-ethylether since it is biomass derived and degraded back to renewable resources (Aycock, 2007; Pace et al., 2012).

6.3.3 Alkoxysilanes and structured silicates from RHA

The derivation of alkoxysilanes from RHA is mentioned briefly in Section 6.2.2, but it is important to detail here. Laine et al., Fukaya et al. and others have set out to use RHA as a source material for the low-temperature catalytic forming of alkoxysilanes, the "Holy Grail" of silicon chemistry (Laine et al. 2016a). The most common methods to acomplish this conversion have been to take acid pretreated RHA, a base catalyst (i.e. NaOH) with diols at atmospheric pressure, or mono-alcohols at increased pressures in a batch synthesis process at temperatures below 300 °C over the course of several hours. Once the alkoxysilanes are formed, they can then be purified by distillation. Table 6.3 shows the maximum total dissolution obtained from a high surface area RHA over the course of 24 h. Though the yields are generally lower than that from pure amorphous fumed silica (surface area 350 m^2/g), these are obtained without first removing the remaining 10–15 % carbon content and are from green sources. Figure 6.16 outlines the possible outcomes from the Laine diol methodology with precipitated/fumed silica and R-alkoxysilanes being major targets (Laine et al., 2014; Furgal & Laine, 2016; Laine et al., 2016a).

Table 6.3: Comparing RHA conversion methods showing alcohols and total % dissolution.

	2-Me-2,4-pentanediol (b.p 197 °C)	Ethanol (b.p.)	Ethylene Glycol (b.p.)
RHA (230 m^2/g)	60 % (Laine et al. 2016a)	71 % (60 % TEOS) (Fukaya et al. 2017)	27 % (Laine et al. 2014)

Since it is possible to make R-trialkoxysilane derivatives directly from RHA derived materials, the next section will explore the synthesis of silsesquioxane and octasilicate based structural motifs.

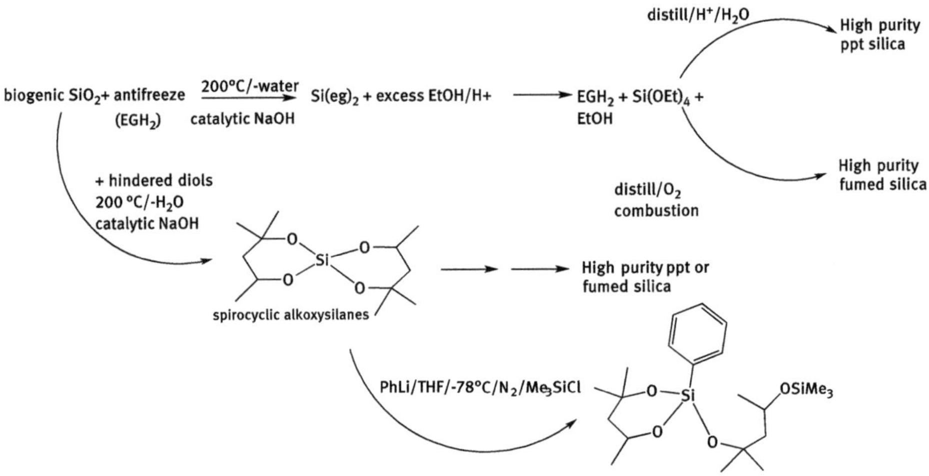

Figure 6.16: General schematic for the conversion of biogenic silica to reactive silicon-based building blocks.

6.3.4 From RHA to silsesquioxane derivatives

Furgal and Laine (2016) also conducted a short study to test the viability of forming silsesquioxane cage compounds directly from the resulting R-alkoxysilanes given in Figure 6.15 using fluoride catalysis (Figure 6.17). The fluoride catalysis (tetrabutylammonium fluoride, TBAF) method worked quite well and resulted in a mixture of full silsesquioxane cage sizes (8,10 and 12 Si) and some partial cages, but gives nearly 100 % conversion of R-alkoxysilanes to cage through the thermodynamic equilibration process (Furgal et al. 2016). Desired cage sizes are separated by selective precipitation with the undesired sizes capable of being recycled back into the reaction sequence and re-equilibrated.

In addition to the direct formation of alkoxysilanes and convential T-type silsesquioxanes (Figure 6.17), Q-type silsesquioxane (nano-structured silicates) derivatives are also possible directly from RHA (Asuncion et al. 2005). They are typically derived from traditional non-biogenic alkoxysilanes such as TEOS (Laine 2005), but Asuncion et al. have developed a procedure (Figure 6.18A) to form octasilicates (octa-anion, OA) using a reaction mixture of RHA, 8-equivalents of ammonium hydroxide and an aqueous alcohol (methanol) (Asuncion et al. 2005). The reaction is relatively slow (20–30 days) with yields maxing out at ~ 50 %; however, it takes place at room temperature with only moderate stirring. The OA can then be worked up after filtration to remove carbon, followed by solvent evaporation to concentrate to generate white crystals and purification by recrystallization from hot methanol. Functionalization of each of the anionic oxygens can be achieved by reaction with any mono-chlorosilanes (i.e. chloro-dimethylsilane) to instill reactive functionalities

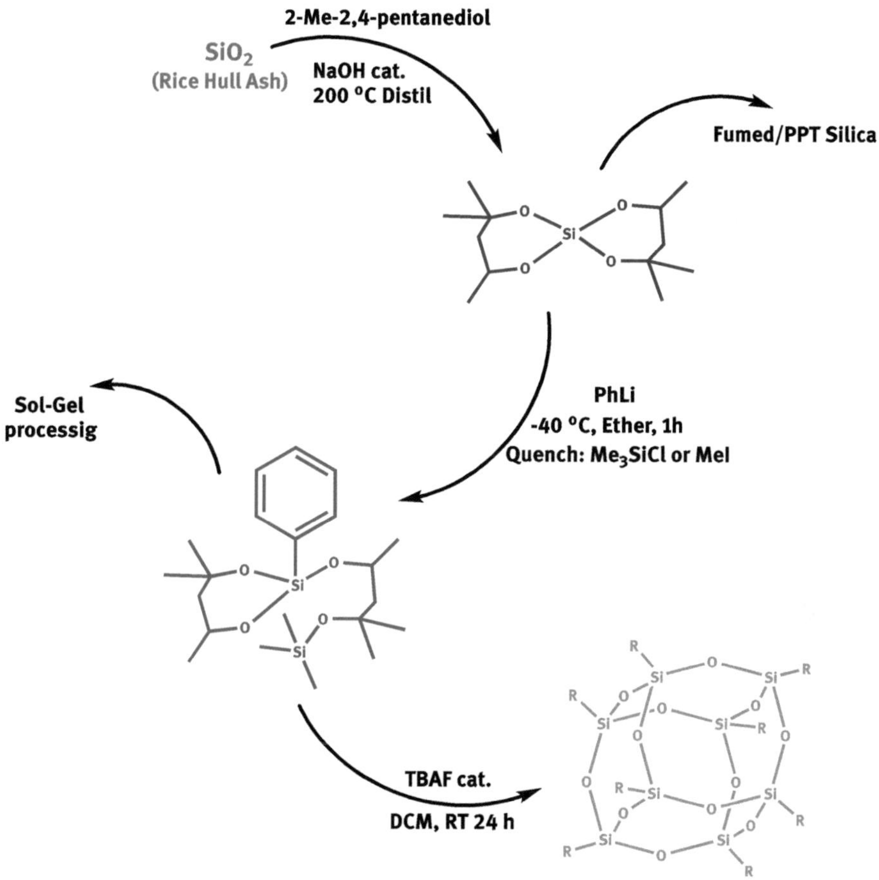

Figure 6.17: Outline of a synthetic process for converting silica to silsesquioxanes (Furgal and Laine 2016).

to make a variety of materials (Figure 6.18B) (Pan et al. 2015). One of the unique features of OA is that it only reacts with silanes and not carbon based electrophiles (Laine 2005).

6.4 Bio-derived siloxanes, composites and other functional silicon-based materials

Experiments involving processing and materials derived from RHA have been the most visible and important breakthroughs of silicon raw material development in recent times; however, there has been little development in polymers such as silicones synthesized from RHA or other green feedstocks. Work by Furgal and Laine

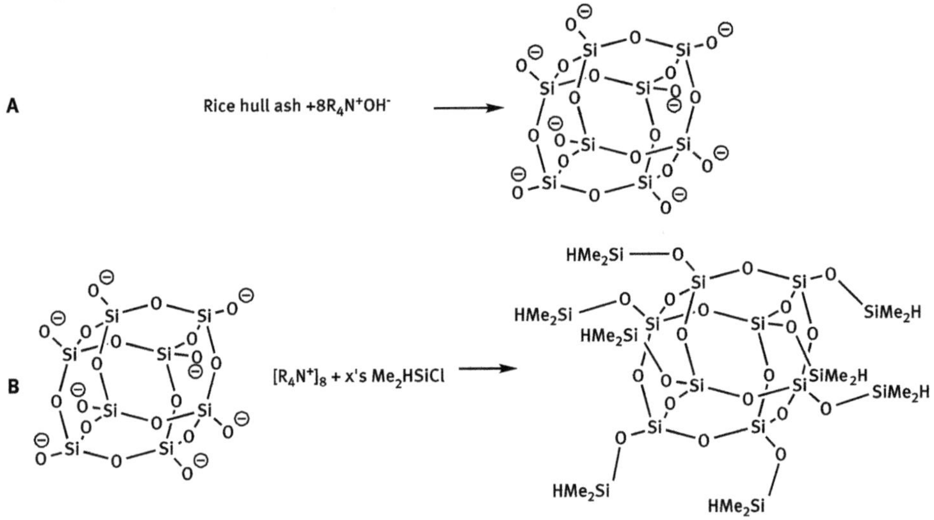

Figure 6.18: (A) Synthesis of octasilicate from RHA; (B) Synthesis of a dimethylsilane functionalized octasilicate (OHS).

(2016) have started the development of R_x-alkoxysilanes derived from RHA; however, only a small study on the synthesis of phenylsilsesquioxane cages has been explored, albeit with successful formation of the desired materials. The reason that more siloxanes have not been explored may be due to the difficulty in forming R_2-alkoxysilanes from RHA derivatives, as Furgal and Laine were able to synthesize the desired alkoxysilanes, but unable to isolate them. So, though it is theoretically possible, it is not yet evident that siloxanes, particularly silicones, have been developed from biogenic and/or other environmentally friendly methods. Therefore, relentless work is still necessary in greening simple silicones for everyday life.

6.4.1 Lignin and gelatin based siloxane copolymers

Even though silicones still demand robust efforts to make them more environmentally friendly on their own, many research groups have explored greening silicones by making bio-polymer and siloxane hybrids in order to reduce the amount of siloxane needed in functional materials.

One area of bio-based siloxane composites being explored is the incorporation of bio-waste materials such as lignin and gelatin (Zhang et al. 2014, 2015a, and 2015b; Macphail and Brook 2017). Lignin, typically from pulping processes (paper/cardboard) can be a very useful polymer for many structural/composite applications; however, it is often difficult to process and/or depolymerize and mostly burned as a cheap fuel. This is

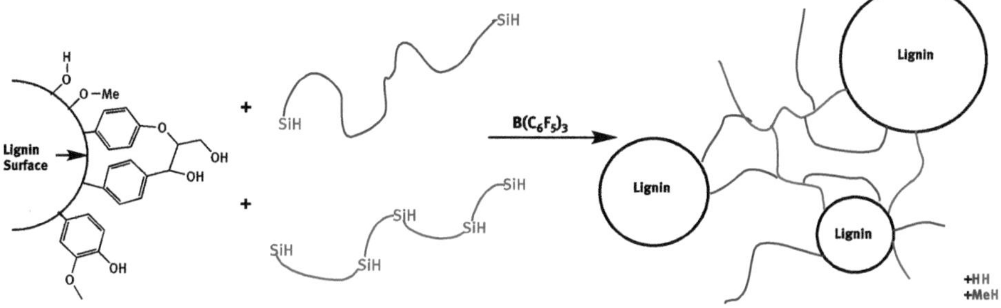

Figure 6.19: Catalyzed oxysilylation of lignin with silane containing siloxane polymers to form copolymers.

especially true of hardwood lignin (Zhang et al. 2015a). Brook et al. have developed a lignin-silicone-based elastomer and subsequent foam composite based on the reaction of lignin with hydride functionalized silicone using the Piers–Rubinsztijn catalysis method with $B(C_6F_5)_3$ (Figure 6.19); and also studied their degradation mechanisms (Zhang et al. 2014, 2015a, and 2015b). They report that in this method Si-H groups on the silicone can be reacted with the alcohol groups present on the surface of lignin, making a copolymer/composite that has similar properties to many standard silicone elastomers. Materials with up to 71 % lignin content were studied, with optimal performance found for samples containing ~41 % lignin by having 150 % break elongations, and modulus approaching 5 MPa. They also found in a subsequent paper that these materials could be foamed by first extrusion followed by injection molding to give mechanical properties on par with other silicone-organic copolymers (Zhang et al. 2015b). Lastly, they also report that the Piers-Rubinsztijn catalyst system can be combined with short chain “monomers” with silane functionality to positively influence depolymerization of soft lignin polymers (Zhang et al. 2014).

Brook et al. have also explored the incorporation of other sacharides/polysacharides into siloxane polymers using the approach discussed above as well as a new approach with boronic acids grafted directly onto the siloxane to remove the need for added boron catalysts (Macphail and Brook 2017). They find that these materials can be loaded and hydrophobic properties adjusted with up to 50 % sorbitol, glucose, pullulan or phytoglycogen nanoparticles and tuned to desired moduli, elasticity and particle sizes.

In addition to lignin, gelatin has also found use in forming copolymers with siloxanes. This area has mostly focused on developing scaffolds for applications as a biomimetic/bioactive in bone tissue engineering (Ren et al. 2001, 2010; Xue et al. 2014). Gelatin itself is a good scaffold material for tissue growth; however, its lack of bioactivity vs. Si-OH groups limits bone development. One approach, developed by Ren et al. (2001) but also used by Xue et al. (2014), uses a sol-gel technique to combine 3-glycidoxypropyltrimethoxysilane with gelatin and Ca^{2+} ions (Ren et al. 2001). After forming the gel structure through NH_4OH catalyzed reactions, the material was then freeze-dried to instill porosity into the structure, controlled by a combination of pH and freezing temperature. In vitro studies with stimulated body fluid showed bioactivity through the formation of apatite. Later work has now developed a microporous mat system made by electrospinning the gelatin-siloxane copolymer, allowing for the introduction of stem cells and improved bioactivity over the bulk material (Ren et al. 2010).

6.4.2 Other organic-siloxane copolymers

Other types of green siloxane copolymers have also been explored and classified either by the materials incorporated or by polymerization technique and have been used for many applications. Yilgor et al. (2003) used isopropanol as an effective "green" solvent for generating urea-rich copolymers of amine-terminated silicones with bis(4-isocyanatocyclohexyl)methane (HMDI). Poojari and Clarson (2010) used lipase catalysis to decrease the environmental impact of the synthesis of aromatic polyester/amide-silicone copolymers. The *lipase candida antarctica* B was used in place of a typical catalyst system and grafted on a porous acrylic support to make alternating copolymers of hydroxy or amine terminated silicones with dimethyl terephthalate. Lipase has also been used by Mosurkal et al. (2011) to synthesize terpolymers containing siloxane, bisimide groups and boron containing units in their backbone. These materials were found to have good flame resistance properties with heat capacities up to 183 $Jg^{-1}K$ and were also coatable onto surfaces. MacIejewski et al. (2009) developed a reusable catalyst system for the hydrosilylation of functionalized olefins in ionic liquids. The rhodium siloxide catalyst can easily perform $R_3Si\text{-}H$, $CH_2=CHR$ couplings in >80 % conversion to $R_3Si\text{-}CH_2CH_2R$ and due to the use of ionic liquids to induce a biphasic system to immobilize the catalyst it has high activity like homogenous catalysis, but is easily recovered and reused like a heterogenous catalyst.

6.4.3 Inorganic copolymers and composites

In addition to siloxane-organic hybrid copolymers, siloxanes can also be combined with inorganic polymers to make environmentally friendly materials and coatings (Tiwari and Hihara 2014). Siloxane ceramic matrix composites (CMCs) can be formed by incorporating RHA directly as a filler material to affect material properties (Siqueira et al. 2009). In this work they pyrolyzed at 1000 and 1600 °C different mixtures of RHA and a pre-ceramic polymer made from hydrosilylation of 1,3,5,7-tetramethyl- 1,3,5,7-tetravinylcyclotetrasiloxane (D4Vi) and poly(methylsiloxane) (PMHS) with a Pt catalyst. After pyrolysis additional siloxane polymer was added to infiltrate the pores and then re-polymerized to induce densification from 1.29 to 1.94 g/cm^3 after two infiltration cycles, with an increase in flexural modulus from 11 to 23 GPa.

Porous inorganic/organic hybrid materials can also be made from green processes. The octasilicates made from RHA and discussed in Section 2.3 are excellent building blocks for porous materials, especially in the form of $[(HMe_2SiOSiO_{1.5})_8]$, (OHS) (Pan et al. 2015). Oxysilylation can be used to react Si-H groups with alkoxysilanes in the presence of $B(C_6F_5)_3$ catalysts. Figure 6.20 shows how this technique is applied to make semi-crystalline microporous materials with surface areas over 700 m^2/g. Due to their high porosity, these materials have potential for gas/solvent storage, insulation and environmental remediation.

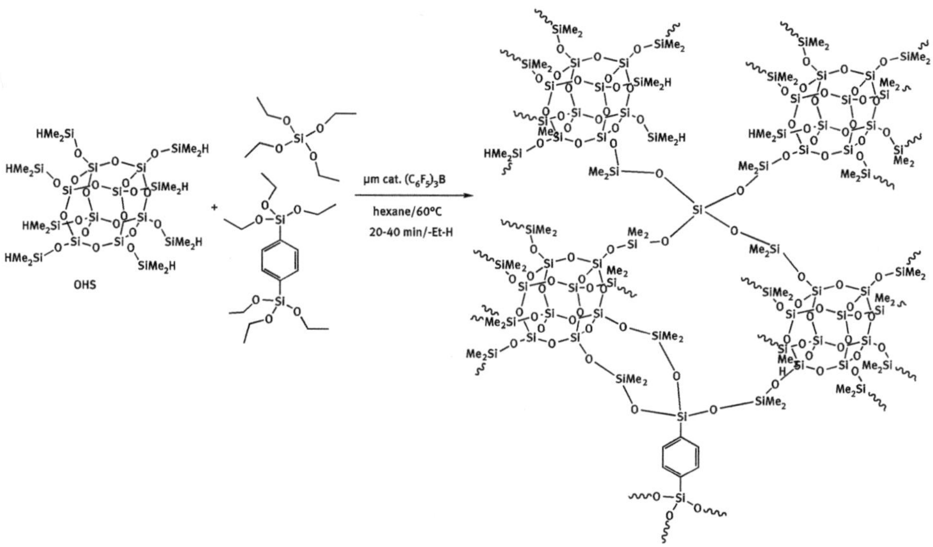

Figure 6.20: Oxysilylation of dimethylsilane functionalized octasilicate with TEOS and 1,4-bis-triethoxysilylbenzene using $B(C_6F_5)_3$ catalyst to form porous networks.

6.5 Degradation pathways and environmental implications

One of the most important concerns in silicon chemistry, especially pertaining to cyclic siloxane and silicone use are their degradation in both nature and within biological systems.

6.5.1 Siloxanes in biological systems

Even though most siloxanes are believed to be inert, their use in implants, medical devices, and contact agents such as make-up have come under strict scrutiny recently, especially in regard to the use/residues of cyclic siloxanes in these applications (Brown et al. 1960; Mojsiewicz-Pienkowska et al. 2016). The European Union has designated the cyclic siloxanes of $D_{3\text{-}6}$ as persistent materials likely to cause biological harm and have ordered their removal from most of these types of uses (Commission Regulation (EU) 2018/1516 2018; ECHA 2018). For instance many lawsuits have been filed against silicone implant makers and implicated cancer caused by the leaching of siloxane agents and platinum catalysts (Schleiter 2010). Though many researchers over the years have sought to either show toxicity or debunk it, no one has given a very thorough review except a comprehensive 50-year literature analysis of silicone toxicology given in the US National Academies Press. This review suggests that silicone in any normal concentration has only nominal toxicological effects in both implanted, ingested, inhaled and cutaneous exposure including small amounts of platinum catalysts and the cyclic siloxanes (Bondurant et al. 1999). For more information on silicone toxicology in biological systems, we direct the readers to that report, which is available as a free download as well as a more recent review by Chao et al. (2016). We also point the reader to a review series on siloxane contact (Mojsiewicz-Pienkowska et al. 2016). From this information, the reader can make their own informed decision about the call for removal of silicones from consumer products. It is likely that it is premature to make the ban on cyclic siloxanes without further studies.

6.5.2 Siloxanes in nature

In nature do siloxanes simply degrade to non-toxic derivatives, or do they cause environmental harm over the long term? In order to understand the above question we must dive into a background study of the degradation mechanisms of siloxane derivatives in many different environments. Many researchers since at least the 1960s have attempted to understand the environmental implications of siloxane degradation under natural conditions, with many early researchers believing that the fully synthetic nature of these materials prohibited them from degrading to safe components in the environment. Degradation may occur in water (Balykova and Rode 1969; Graiver et al. 2003), soil (Lehmann et al., 1998; Xu et al., 1998; Grumping et al., 1999; Lehmann et al. 2000; Varaprath and Larson 2002), atmospheric (Corning 1998; Graiver et al. 2003), and UV oxidative (Frye et al. 1984) conditions amongst many other more complex environments. Here we give a brief overview of the natural degradation pathways of siloxanes and suggest a detailed review by Narayan et al. for an in-depth analysis of the natural degradation pathways of silicones (Graiver et al. 2003).

In general, silicones do not degrade quickly in the environment, since to gain effective hydrolysis, the environment must be either strongly acidic or basic. However degradation does take place slowly over time through hydrolysis with water (at either slightly acidic or basic pH) and can subsequently convert the siloxane backbone to dimethylsilanediol followed by even slower oxidation of the methyl groups to aldehyde and then CO_2 by bacterial degradation with *Arthobacter* and *Fusarium oxysporium schlechtendahl* (Stevens, 1998; Grumping et al., 1999;

Graiver et al. 2003). Though this process is relatively straightforward, it has been found in competition with a process that is 2–20 times faster. The more likely degradation scenario in both aqueous and soil environments relies on the hydrolysis of siloxanes to low molecular weight and volatile oligomeric R_2-silanols or cyclic siloxanes ($D_{3\text{-}6}$), which then evaporate into the atmosphere. Note: In soil most high molecular weight materials must undergo extensive depolymerization first, which takes place fastest (4–7 days) in dry, high sand content soils. It is then in the atmosphere that the most effective degradation occurs by oxidation with UV light, ozone, nitrogen trioxide radicals or hydroxyl radicals (Frye et al. 1984; Graiver et al. 2003). Figure 6.21 shows the proposed pathway for oxidation of the Si-CH_3 functional group in the atmosphere using hydroxyl radical, oxygen and peroxide. As an example, reaction rates are on the order of 10^{-21} in O_3, 10^{-17} in $^{\cdot}NO_3$, and 10^{-12} $cm^3mol^{-1}sec^{-1}$ in $^{\cdot}OH$, suggesting that the fastest reactions take place with hydroxyl radicals. The reaction pathway is quite complex and relies on the formation of many different radical species for full oxidation to airborne silica and CO_2. It is expected that the degradation products would have minimal, if any, adverse effects on terrestrial/aquatic organisms or plants and that the degradation of these materials in the atmosphere have no effect on overall air quality and pose little to no environmental risk.

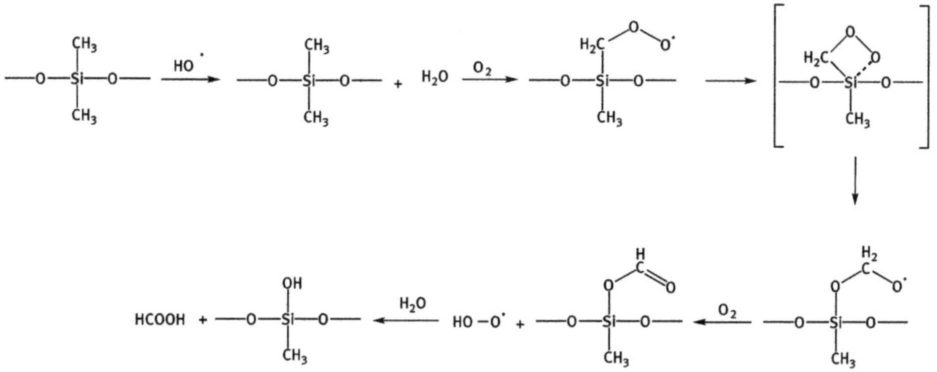

Figure 6.21: Atmospheric degradation pathway of dimethyldioxysilanes with hydroxyl radical (Graiver et al. 2003).

6.6 Conclusions

The greening of silicon chemistry is a slow and tedious effort that has made great strides over the years toward limiting environmental impact; however, there is much more to do. We find that RHA and other biogenic silica sources are the most likely raw materials moving forward toward the development of improved synthesis methods for functional silicon-based materials including alkoxysilanes, silicon, silica and siloxanes. One of the distinct advantages of silicon-based materials is their true versatility; they can be used on their own as sealants, adhesives and coatings as well as combined with many biological materials for composites and copolymers, used as ceramics (insulators), and semi-conductors. Most important of all, their subsequent degradation has minimal negative environmental impact. Silicon-based materials hold an important place in our developed world and will be important for years to come. Therefore, we call on all silicon chemists to think about the environmental impact of their materials and develop new strategies to make their lifecycle from synthesis to end of use more cradle to cradle and less cradle to grave.

References

Abe, Y. & Gunji, T. 2004, 'Oligo- and polysiloxanes', *Progress in Polymer Science*, vol. 29, pp.149–182, 10.1016/j.progpolymsci.2003.08.003.

Adam, F., Appaturi, J. N. & Iqbal, A. 2012, 'The utilization of rice husk silica as a catalyst: Review and recent progress', *Catalysis Today*, Elsevier B.V., vol. 190, pp.2–14, 10.1016/j.cattod.2012.04.056.

Adams, F. V. & Mulaba-Bafubiandi, A. F. 2014, 'Application of rice hull ash for turbidity removal from water', *Physics and Chemistry of the Earth*, 10.1016/j.pce.2014.09.010.

Akiyama, M., Suzuki, E. & Ono, Y. 1993, 'Direct synthesis of tetramethoxysilane from rice hull ash by reaction with dimethyl carbonate', *Inorganica Chimica Acta*, vol. 207, pp.259–261, 10.3839/jksabc.2.

An, D., Guo, Y., Zhu, Y. & Wang, Z. 2010, 'A green route to preparation of silica powders with rice husk ash and waste gas', *Chemical Engineering Journal*, Elsevier B.V., vol. 162, pp.509–514, 10.1016/j.cej.2010.05.052.

Andrianov, K. A. 1965, *Metalorganic polymers*, Interscience, New York.

Arkles, B., Steinmetz, J., Zazyczny, J. & Mehta, P. 1992, 'Factors contributing to the stability of alkoxysilanes in aqueous solution', *Journal of Adhesion Science and Technology*, pp.193, 10.1163/156856192X00133.

Arkles, B. 1996, 'Gringard reagents and silanes', in G. Silverman & P. Rakita (eds), *Handbook of grignard reagents 1*, New York, Marcel Dekker, pp.533–538, 667–672.

Asimov, I. 1964, *Asimov's biographical encyclopedia of science and technology*, Doubleday & Company, Garden City, NY.

Asuncion, M. Z., Hasegawa, I., Kampf, J. W. & Laine, R. M. 2005, 'The selective dissolution of rice hull ash to form $[OSiO_{1.5}]_8[R_4N]_8$(R = Me, CH_2CH_2OH) octasilicates. Basic nanobuilding blocks and possible models of intermediates formed during biosilicification processes', *Journal of Materials Chemistry*, vol. 15, pp.2114, 10.1039/b502178b.

Aycock, D. F. 2007, 'Solvent applications of 2-methyltetrahydrofuran in organometallic and biphasic reactions', *Organic Process Research and Development*, vol. 11, pp.156–159, 10.1021/op060155c.

Bailey, R. & O'Connor, F. 1959, 'Preparation of Alkyl Orthosilicates by Reacting Silicon Dioxide with an Alcohol'. United States Patent Office.

Bakar, R. A., Yahya, R. & Gan, S. N. 2016, 'Production of high purity amorphous silica from rice husk', *Procedia Chemistry*, Elsevier Ltd., vol. 19, pp.189–195, 10.1016/j.proche.2016.03.092.

Balykova, T. N. & Rode, V. V. 1969, 'Progress in the study of the degradation and stabilisation of siloxane polymers', *Russian Chemical Reviews*, vol. 38, pp.306–317, 10.1070/RC1969v038n04ABEH001743.

Barnum, D. 1970, 'Catechol complexes with silicon', *Inorganic Chemistry*, vol. 9, pp.1942–1943, 10.1021/ic50090a038.

Barnum, D. 1972, 'Reaction of catechol with colloidal silica and silicic acid in aqueous ammonia', *Inorganic Chemistry*, vol. 11, pp.1424–1429, http://pubs.acs.org/doi/abs/10.1021/ic50112a052.

Blohowiak, K. Y., Treadwell, D. R., Mueller, B. L., Hoppe, M. L., Jouppi, S., Pansal, P., Chew, K. W., Scotto, C. L. S., Babonneau, F., Kampf, J. & Laine, R. M. 1994, 'Sio_2 as a starting material for the synthesis of pentacoordinate silicon complexes .1', *Chemistry of Materials*, vol. 6, pp.2177–2192, 10.1021/cm00047a045.

Boer, F. P., Flynn, J. J. & Turley, J. W. 1968, 'Structural studies of pentacoordinate silicon. III. Tetramethylammonium bis(o-phenylenedioxy)phenylsiliconate', *Journal of the American Chemical Society*, vol. 90, pp.6973–6977, 10.1021/ja01027a014.

Bondurant, S., Ernster, V. & Herdman, R. (eds.) 1999, 'Silicone toxicology', in *Institute of medicine (Us) committee on the safety of silicone breast implants*, National Acadamies Press (US), Washington, D.C., pp.80–113, 10.17226/9602.

Bonitz, E. & Pfalz, F. 1970, 'Production of Organosilsanes'. United States Patent Office.

Boudin, A, Cerveau, G. & Chuit, C. 1988a, 'Reactivity of dianionic hexacoordinated silicon complexes toward nucleophiles: A new route to organosilanes from silica', *Organometallics*, vol. 7, pp.1165–1171, 10.1021/om00095a023.

Boudin, A., Cerveau, G., Chuit, C., Corriu, R. J. P. & Reye, C. 1986, 'Reaction of grignard reagents with dianionic hexacoordinated silicon complexes: Organosilicon compounds from silica gel', *Angewandte Chemie International Edition in English*, vol. 25, pp.474–476, 10.1002/anie.198604741.

Boudin, A., Cerveau, G., Chuit, C., Corriu, R. J. P. & Reye, C. 1988b, 'Reactivity of hypervalent species: Reactions of anionic penta-coordinated silicon complexes towards nucleophiles.', *Bulletin of the Chemical Society of Japan*, pp.101–106, 10.1246/bcsj.61.101.

Brown, J. B., Ohlwiler, D. A. & Fryer, M. P. 1960, 'Investigation of, and use of dimethyl siloxanes, halogenated carbons and polyvinyl alcohol, as subcutaneous prosthesis *', *Annals of surgery*, pp.534–547, 10.1097/00000658-196015230-00015.

Butera, V., Fukaya, N., Choi, J. C., Sato, K. & Choe, Y. K. 2018, 'Alkoxysilane production from silica and dimethylcarbonate catalyzed by alkali bases: A quantum chemical investigation of the reaction mechanism', *Inorganica Chimica Acta, Elsevier*, vol. 482, pp.70–76, 10.1016/j.ica.2018.05.036.

Chandrasekhar, S., Satyanarayana, K. G., Pramada, P. N. & Raghavan, P. 2003, 'Review processing, properties and applications silica from rice husk', *Journal of Materials Science*, vol. 38, pp.3159–3168.

Chao, A. H., Garza, R. & Povoski, S. P. 2016, 'A review of the use of silicone implants in breast surgery', *Expert Review of Medical Devices*, Taylor & Francis, vol. 13, pp.143–156, 10.1586/17434440.2016.1134310.

Cheng, H., Tamaki, R., Laine, R. M., Babonneau, F., Chujo, Y. & Treadwell, D. R. 2000, 'Neutral alkoxysilanes from silica', *Journal of the American Chemical Society*, vol. 122, pp.10063–10072, 10.1021/ja001885h.

Colas, A. & Curtis, J. 2004, 'An introduction to materials in medicine', *Silicone biomaterials: History and chemistry*, pp.80–85, 698-707, https://www.dowcorning.com.cn/zh_CN/content/publishedlit/52-1069-01.pdf%5Cn, http://sutlib2.sut.ac.th/sut_contents/H87104.pdf.

Commission Regulation (EU) 2018/1516. 2018, *European Commission*, European Union.

Corning, D. 1998, *Degradation of silicone polymers in nature*, 10.1109/CEE-SECR.2009.5501173.

Dosaj, V., Kroupa, M. & Bittar, R. 2000, 'Silicon and silicon alloys, chemical and metallurgical', in *Kirk-Othmer encyclopedia of chemical technology*, John Wiley & Sons, Inc., pp.1–25, 10.1002/0471238961.0308051304151901.a01.pub2.

ECHA. 2018, *Annex XVII to Reach – Conditions of restriction, European Union*, https://echa.europa.eu/documents/10162/50e79685-efaf-ac9a-4acb-d8be3f0e9ddc.

Foletto, E. L., Gratieri, E., de Oliveira, L. H. & Jahn, S. L. 2006, 'Conversion of rice hull ash into soluble sodium silicate', *Materials Research*, vol. 9, pp.335–338, 10.1590/S1516-14392006000300014.

Falcone, J. 2007, 'Silicon compounds: Anthropogenic silicas and silicates', in *Kirk-Othmer encyclopedia of chemical tech*, John Wiley & Sons, Inc., pp.365–547, 10.1002/0471238961.1925142006011203.a01.pub22005.

Fapohunda, C., Akinbile, B. & Shittu, A. 2017, 'Structure and properties of mortar and concrete with rice husk ash as partial replacement of ordinary Portland cement – A review', *International Journal of Sustainable Built Environment*, The Gulf Organisation for Research and Development, vol. 6, pp.675–692, 10.1016/j.ijsbe.2017.07.004.

Flynn, J. J. & Boer, F. P. 1969, 'Structural studies of hexacoordinate silicon. Tris(o-phenylenedioxy) siliconate', *Journal of the American Chemical Society*, vol. 91, pp.5756–5761, 10.1021/ja01049a010.

Frye, C. L. 1964, 'Pentacoordinate silicon derivatives. II. 1 salts of Bis (o-arylenedioxy) organosiliconic acids', *Journal of the American Chemical Society*, vol. 86, pp.3170–3171, 10.1021/ja01069a054.

Frye, C. L. 1969, 'Stable silicon heterocyclic derivatives of branched alkane-diols', *The Journal of Organic Chemistry*, vol. 34, pp.2496–2499, 10.1021/jo01261a002.

Frye, C. L., Annelin, R. B., Lane, T. H. & Buch, R. R. 1984, 'Photolytic oxidative demethylation of aqueous dimethylsiloxanols', *Environmental Toxicology and Chemistry*, vol. 3, pp.215–222.

Fukaya, N., Choi, S. J., Horikoshi, T., Kataoka, S., Endo, A., Kumai, H., Hasegawa, M., Sato, K. & Choi, J.-C. 2017a, 'Direct synthesis of tetraalkoxysilanes from silica and alcohols', *New Journal of Chemistry*, Royal Society of Chemistry, vol. 41, pp.2224–2226, 10.1039/C6NJ03897B.

Fukaya, N., Choi, S. J., Horikoshi, T., Kataoka, S., Endo, A., Kumai, H., Hasegawa, M., Sato, K. & Choi, J. C. 2017b, 'Direct synthesis of tetraalkoxysilanes from silica and alcohols', *New Journal of Chemistry*, Royal Society of Chemistry, vol. 41, pp.2224–2226, 10.1039/c6nj03897b.

Fukaya, N., Choi, S. J., Horikoshi, T., Kumai, H., Hasegawa, M., Yasuda, H., Sato, K. & Choi, J.-C. 2016, 'Synthesis of tetramethoxysilane from silica and methanol using carbon dioxide and an organic dehydrating reagent', *Chemistry Letters*, vol. 45, pp.828–830, 10.1246/cl.160329.

Furgal, J. C., Goodson III, T. & Laine, R. M. 2016, 'D_{5h} $[PhSiO_{1.5}]_{10}$ synthesis via F^- catalyzed rearrangement of $[PhSiO_{1.5}]_n$. An experimental/computational analysis of likely reaction pathways', *Dalton Transactions*, Royal Society of Chemistry, vol. 45, pp.1025–1039, 10.1039/C5DT04182A.

Furgal, J. C., Jung, J. H., Goodson, T. & Laine, R. M. 2013, 'Analyzing structure-photophysical property relationships for isolated T_8, T_{10}, and T_{12} stilbenevinylsilsesquioxanes', *Journal of the American Chemical Society*, vol. 135, pp.12259–12269, 10.1021/ja4043092.

Furgal, J. C. & Laine, R. M. 2016, 'Nucleophilic attack of R-lithium at tetrahedral silicon in alkoxysilanes. An alternate mechanism', *Bulletin of the Chemical Society of Japan*, vol. 89, pp.705–725, 10.1246/bcsj.20160039.

George, P. D., Sommer, L. H. & Whitmore, F. C. 1955, 'Steric effects in grignard couplings with alkoxysilanes 1', *Journal of the American Chemical Society*, vol. 77, pp.6647–6649, 10.1021/ja01629a079.

Ghofrani, M., Mokaram, K. N., Ashori, A. & Torkaman, J. 2015, 'Fiber-cement composite using rice stalk fiber and rice husk ash: Mechanical and physical properties', *Journal of Composite Materials*, vol. 49, pp.3317–3322, 10.1177/0021998314561813.

Goodwin, G. B. & Kenney, M. E. 1990, 'A new route to alkoxysilanes and alkoxysiloxanes of use for the preparation of ceramics by the sol-gel technique', *Inorganic Chemistry*, vol. 29, pp.1216–1220, 10.1021/ic00331a021.

Goodyear Reaches Supply Agreements For Rice Husk Ash Silica. 2015, https://corporate.goodyear.com/en-US/media/news/goodyear-reaches-supply-agreements.html.

Graiver, D., Farminer, K. W. & Narayan, R. 2003, 'A Review of the fate and effects of silicones in the environment', *Journal of Polymers and the Environment*, vol. 11, pp.129–136, 10.1023/A:1026056129717.

Grumping, R., Michalke, K., Hirner, A. V. & Hensel, R. 1999, 'Microbial degradation of octamethylcyclotetrasiloxane', *Applied and Environmental Microbiology*, vol. 65, pp.2276–2278.

Gurav, A., Kodas, T., Pluym, T. & Xiong, Y. 1993, 'Aerosol processing of materials', *Aerosol Science and Technology*, pp.411–452, 10.1080/02786829308959650.

Hamad, M. A. & Khattab, I. A. 1981, 'Effect of the combustion process on the structure of rice hull silica', *Thermochimica Acta*, vol. 48, pp.343–349, 10.1016/0040-6031(81)80255-9.

Hazimeh, H., Mattalia, J. M., Attolini, M., Bodineau, N., Marchi-Delapierre, K. H. C., Péralez, E. & Chanon, M. 2006, 'Grignard reagent and green chemistry: Mechanistic studies to understand the molecular origins of selectivity in the formation of RMgX', *Indian Journal of Chemistry - Section B Organic and Medicinal Chemistry*, vol. 45, pp.2270–2280.

Huang, Y., Liu, T., Wang, C. & Wang, J. 2013, 'Mechanism and kinetics of the synthesis of phenyltrichlorosilane from trichlorosilane and chlorobenzene by gas phase condensation', *Chemical Engineering Journal*, vol. 226, pp.255–262, 10.1016/j.cej.2013.04.046.

Hunt, L. P., Dismukes, J. P. & Amick, J A. 1982, 'Rice hulls as a raw material for producing silicon', *Journal of the Electrochemical Society*, vol. 131, pp.5–8, 10.1149/1.2115937.

Ikram, N. & Akhter, M. 1988, 'X-ray diffraction analysis of silicon prepared from rice husk ash', *Journal of Materials Science*, vol. 23, pp.2379–2381, 10.1007/BF01111891.

Innovations of Silicone. 2015. viewed 1 January 2015, http://www.dowcorning.com/content/discover/discoverinnovate/.

Jean, P. A., Gallavan, R. H., Kolesar, G. B., Siddiqui, W. H., Oxley, J. A. & Meeks, R. G. 2006, 'Chlorosilane acute inhalation toxicity and development of an LC50 prediction model', *Inhalation Toxicology*, vol. 18, pp.515–522, DOI: M281460710597351 [pii]\r10.1080/08958370600748695.

Jung, J. H., Furgal, J., Iii, T. G., Mizumo, T., Schwartz, M., Chou, K., Vonet, J.-F. & Laine, R. M. 2012, '3-D molecular mixtures of catalytically functionalized $[vinylSiO_{1.5}]_{10}/[vinylSiO_{1.5}]_{12}$. photophysical characterization of second generation derivatives', *Chemistry of Materials*, vol. 24, pp.1883–1895. viewed 8 January 2013, http://pubs.acs.org/doi/abs/10.1021/cm300587s.

Jung, J. H. & Laine, R. M. 2011, 'Beads on a chain (BOC) polymers formed from the reaction of mixtures (x = 2-4) with the diglycidyl ether of bisphenol A', *Macromolecules*, vol. 44, pp.7263–7272.

Jung, K.-H., Kim, S.-Y., Tan, W., Shin, D.-S. & Ahn, C. 2005, 'Improved methods for the synthesis of aryltriealkoxysilane derivatives via aryllithium and arylGrignard bromide with tetra-alkoxysilane', *Bulletin Institute for Basic Science*, vol. 17, pp.79.

Kalapathy, U., Proctor, A. & Shultz, J. 2003, 'Silicate thermal insulation material from rice hull ash', *Industrial and Engineering Chemistry Research*, vol. 42, pp.46–49, 10.1021/ie0203227.

Kamath, S. R. & Proctor, A. 1998, 'Silica gel from rice hull ash: Preparation and characterization', *Cereal Chemistry*, 10.1094/CCHEM.1998.75.4.484.

Kapur, P. 1985, 'Production of reactive bio-silica from the combustion of rice husk in a tube-in-basket (TiB) burner', *Powder Technology*, vol. 44, pp.63–67, 10.1016/0032-5910(85)85022-1.

Kawamoto, H. & Okuwada, K. 2007, 'Development trend for high purity silicon raw material technologies— Expecting innovative silicon manufacturing processes for solar cell applications', *Science and Technology Trends*, vol. 24, pp.38.

Kenney, M. E. & Goodwin, G. B. 1988, 'Silicate Esters and Organosilicon Compounds'. United States Patent Office. doi: 10.1103/PhysRevE.87.021001.

Kinrade, S. D., Del Nin, J. W., Schach, A S., Sloan, T. A, Wilson, K. L. & Knight, C. T. 1999, 'Stable five- and six-coordinated silicate anions in aqueous solution', *Science* (New York, N.Y.), vol. 285, pp.1542–1545, 10.1126/science.285.5433.1542.

Klokov, B. A. 2001, 'Continuous and batch organomagnesium synthesis of ethyl-substituted silanes from ethylchloride, tetraethoxysilane, and organotrichlorosilane for production of polyethylsiloxane liquids. 2. Continuous one-step synthesis of ethylethoxy- and ethylchlorosilan', *Organic Process Research and Development*, vol. 5, pp.234–240, 10.1021/op000100p.

Klosowski, J. M. & Wolf, A. T. 2009, 'The history of sealants', in K. L. Mittal & A. Pizzi (eds.), *Handbook of sealant technology*, CRC press, Boca Raton, FL, pp.1–22.

Krishnarao, R. V. 1992a, 'Distribution of silica in rice husks and its effect on the formation of silicon carbide', *Ceramics International*, vol. 18, pp.243–9.

Krishnarao, R. V. 1992b, 'Effect of Si_3N_4 additions on the format of SiC whiskers from rice Husks', *Ceramics International,* vol. 18, pp.185–191.

Krishnarao, R. V., Mahajan, Y. R. & Kumar, T. J. 1998, 'Conversion of raw rice husks to sic by pyrolysis in nitrogen atmosphere', *Journal of the European Ceramic Society*, vol. 18, pp.147–152, 10.1016/S0955-2219(97)00093-9.

Krishnarao, R. V. 1992a, Distribution of silica in rice husks and its effect on the formation of silicon carbide, *Ceramics International*, vol. 18, pp.243–249.

Krishnarao, R. V. 1992b, 'Effect of Si_3N_4 Additions on the format, o-nofSiC whiskers from rice H usks', *Ceramics International*, vol. 18, pp. 185–191.

Laine, R. M. 2005, 'Nanobuilding blocks based on the $[OSiO_{1.5}]_x$ (x= 6, 8, 10) octasilsesquioxanes', *Journal of Materials Chemistry*, vol. 15, pp.3725, 10.1039/b506815k.

Laine, R. M., Blohowiak, K. Y., Robinson, T. R., Hoppe, M. L., Nardi, P., Kampf, J. & Uhm, J. 1991, 'Synthesis of pentacoordinate silicon complexes from SiO_2', *Nature*, vol. 353, pp.642–644, 10.1038/353642a0.

Laine, R. M., Furgal, J. C., Doan, P., Pan, D., Popova, V. & Zhang, X. 2016a, 'Avoiding carbothermal reduction: Distillation of alkoxysilanes from biogenic, green, and sustainable sources', *Angewandte Chemie International Edition*, vol. 55, pp.1065–1069, 10.1002/anie.201506838.

Laine, R. M., Furgal, J. C., Popova, V. & Yi, E. 2016b, 'Synthesis Method of Alkoxysilanes'. United States: United States Patent Office.

Laine, R. M., Krug, D. J., Marchal, J. C. & McColm, A. 2013, 'Low cost routes to high purity silicon and derivatives thereof'.

Laine, R. M., Marchal, J. C., Popova, V. & Krug, D. J. 2014, 'Method of Producing Alkoxysilanes and Precipitated Silicas from Biogenic Silicas'. United States Patent Office.

Laine, R. M. & Roll, M. F. 2011, 'Polyhedral phenylsilsesquioxanes', *Macromolecules*, vol. 44, pp.1073–1109, 10.1021/ma102360t.

Lee, J. G. & Cutler, I. B. 1975, 'Formation of silicon carbide from rice hulls', *American Ceramic Society Bulletin*, 10.1016/S0008-4433(97)00017-7.

Lee, J. G., Miller, P. D. & Cutler, I. B. 1977, 'Carbothermal reduction of silica bt - reactivity of solids', in J. Wood, O. Lindqvist, C. Helgesson & N.-G. Vannerberg (eds.), *Reactivity of solids*, Springer US, Boston, MA, pp.707–711, 10.1007/978-1-4684-2340-2_102.

Lehmann, R. G., Miller, J. R. & Kozerski, G. E. 2000, 'Degradation of silicone polymer in a field soil under natural conditions', *Chemosphere*, vol. 41, pp.743–749, 10.1016/S0045-6535(99)00430-0.

Lehmann, R. G., Miller, J. R., Xu, S., Singh, U. B. & Reece, C. F. 1998, 'Degradation of silicone polymer at different soil moistures', *Environmental Science and Technology*, vol. 32, pp.1260–1264, 10.1021/es970918z.

Li, J., Shirai, T. & Fuji, M. 2013, 'Rapid carbothermal synthesis of nanostructured silicon carbide particles and whiskers from rice husk by microwave heating method', *Advanced Powder Technology*, The Society of Powder Technology Japan, vol. 24, pp.838–843, 10.1016/j.apt.2013.02.003.

Liou, T. H. 2004, 'Preparation and characterization of nano-structured silica from rice husk', *Materials Science and Engineering A*, vol. 364, pp.313–323, 10.1016/j.msea.2003.08.045.

Liu, N., Huo, K., McDowell, M. T., Zhao, J. & Cui, Y. 2013, 'Rice husks as a sustainable source of nanostructured silicon for high performance Li-ion battery anodes', *Scientific Reports*, vol. 3, pp.1–7, 10.1038/srep01919.

Ma, X., Zhou, B., Gao, W., Qu, Y., Wang, L., Wang, Z. & Zhu, Y. 2012, 'A recyclable method for production of pure silica from rice hull ash', *Powder Technology*, Elsevier B.V., vol. 217, pp.497–501, 10.1016/j.powtec.2011.11.009.

Maclejewski, H., Szubert, K., Marciniec, B. & Pernak, J. 2009, 'Hydrosilylation of functionalised olefins catalysed by rhodium siloxide complexes in ionic liquids', *Green Chemistry*, vol. 11, pp.1045–1051, 10.1039/b819310j.

Macphail, B. & Brook, M. A. 2017, 'Controlling silicone-saccharide interfaces: Greening silicones', *Green Chemistry*, Royal Society of Chemistry, vol. 19, pp.4373–4379, 10.1039/c7gc02088k.

Manoso, A. S., Ahn, C., Soheili, A., Handy, C. J., Correia, R., Seganish, W. M. & DeShong, P. 2004, 'Improved synthesis of aryltrialkoxysilanes via treatment of aryl Grignard or lithium reagents with tetraalkyl orthosilicates', *Journal of Organic Chemistry*, vol. 69, pp.8305–8314, 10.1021/jo048667h.

Marchal, J. C., Krug, D. J., McDonnell, P., Sun, K. & Laine, R. M. 2015, 'A low cost, low energy route to solar grade silicon from rice hull ash (RHA), a sustainable source', *Green Chemistry*, Royal Society of Chemistry, vol. 17, pp.3931–3940, 10.1039/c5gc00622h.

McBee, E. T., Roberts, C. W., Judd, G. F. & Chao, T. S. 1955, 'Reaction of grignard reagents with alkoxysilanes at elevated temperatures', *Journal of the American Chemical Society*, vol. 77, pp.1292–1293, 10.1021/ja01610a066.

Mizuki, E., Okumura, S., Saito, H. & Murao, S. 1993, 'Formation of silicon carbide from rice husks using enzymatic methods for carbon control', *Bioresource Technology*, 10.1016/0960-8524(93)90206-Q.

Mojsiewicz-Pienkowska, K., Jamrógiewicz, M., Szymkowska, K. & Krenczkowska, D. 2016, 'Direct human contact with siloxanes (silicones) - safety or risk part 1. Characteristics of siloxanes (silicones)', *Frontiers in Pharmacology*, vol. 7, pp.1–8, 10.3389/fphar.2016.00132.

Mosurkal, R., Kirby, R., Muller, W. S., Soares, J. W. & Kumar, J. 2011, 'Simple green synthesis of polyborosiloxanes as environmentally-safe, non-halogenated flame retardant polymers', *Green Chemistry*, vol. 13, pp.659–665, 10.1039/c0gc00376j.

Niyomwas, S. 2009, 'Synthesis and characterization of silicon-silicon carbide composites from rice husk ash via self-propagating high temperature synthesis', *Journal of Metals, Materials and Minerals*, vol. 19, pp.21–25.

Okamoto, M. 1994, 'Reaction pathway of formation of methoxysilanes in the reaction of silicon with methanol catalyzed by copper(i) chloride', *Journal of Catalysis*, vol. 145, pp.537–543, 10.1006/jcat.1994.1064.

Okutani, T. 2009, 'Utilization of silica in rice hulls as raw materials for silicon semiconductors', *Journal of Metals, Materials and Minerals*, vol. 19, pp.51–59.

Pace, V., Hoyos, P., Castoldi, L., Domínguez De María, P. & Alcántara, A. R. 2012, '2-Methyltetrahydrofuran (2-MeTHF): A biomass-derived solvent with broad application in organic chemistry', *ChemSusChem*, vol. 5, pp.1369–1379, 10.1002/cssc.201100780.

Palma-Barrera, J. P., Sánchez-Ramírez, E., Ramírez-, C., Cervantes-Jauregui, J. A., Gabriel, J. & Hernández, S. 2018, 'Reactive distillation column design for tetraethoxysilane (teos) production. Part Ii : Dynamic properties and inherent safety', *Industrial and Engineering Chemistry Research*, ASAP, 10.1021/acs.iecr.8b04962.

Pan, D., Yi, E., Doan, P. H., Furgal, J. C., Schwartz, M., Clark, S., Goodson Iii, T. & Laine, R. M. 2015, 'Microporous inorganic/organic hybrids via oxysilylation of a cubic symmetry nanobuilding block [$(HMe_2SiOSiO_{1.5})_8$] with $R_xSi(OEt)_{4-x}$', *Journal of the Ceramic Society of Japan*, vol. 123, pp.756–763, 10.2109/jcersj2.123.756.

Pode, R. 2016, 'Potential applications of rice husk ash waste from rice husk biomass power plant', *Renewable and Sustainable Energy Reviews*, Elsevier, vol. 53, pp.1468–1485, 10.1016/j.rser.2015.09.051.

Poojari, Y. & Clarson, S. J. 2010, 'Lipase-catalyzed synthesis and properties of silicone aromatic polyesters and silicone aromatic polyamides', *Macromolecules*, vol. 43, pp.4616–4622, 10.1021/ma100548z.

Proctor, A., Shultz, J. & Kalapathy, U. 2000, 'A simple method for production of pure silica from rice hull ash', *Bioresource Technology*, vol. 73, pp.257–262, 10.1016/S0960-8524(99)00127-3.

Quispe, I., Navia, R. & Kahhat, R. 2017, 'Energy potential from rice husk through direct combustion and fast pyrolysis: A review', *Waste Management*, Elsevier Ltd, vol. 59, pp.200–210, 10.1016/j.wasman.2016.10.001.

Reisch, M. S. 2011, 'Storm over silicones', *Chemical & Engineering News*, vol. 89, pp.10–13, DOI:10.1021/CEN042611142704.

Reisch, M. S. 2012, *Canada clears siloxane D5, Chemical & Engineering News*.

Ren, L., Tsuru, K., Hayakawa, S. & Osaka, A. 2001, 'Sol-gel preparation and in vitro deposition of apatite on porous gelatin-siloxane hybrids', *Journal of Non-Crystalline Solids*, vol. 285, pp.116–122, 10.1016/S0022-3093(01)00441-0.

Ren, L., Wang, J., Yang, F. Y., Wang, L., Wang, D., Wang, T. X. & Tian, M. M. 2010, 'Fabrication of gelatin-siloxane fibrous mats via sol-gel and electrospinning procedure and its application for bone tissue engineering', *Materials Science and Engineering C*, vol. 30, pp.437–444, 10.1016/j.msec.2009.12.013.

Rodrigues, F. A. 2003, 'Low-temperature synthesis of cements from rice hull ash', *Cement and Concrete Research*, 10.1016/S0008-8846(03)00104-2.

Rosenberg, J. (no date), *The history of silly putty*, viewed: 1 January 2015, http://history1900s.about.com/od/1950s/a/sillyputty.htm.

Rosenheim, A., Raibmann, B. & Schendel, G. 1931, 'Über innerkomplexe Brenzcatchinate vierwertiger Elemente', *Zeitschrift fur Anorganische Chemie*, vol. 196, pp.160–176.

Rozainee, M., Ngo, S. P., Salema, A. A. & Tan, K. G. 2008a, 'Fluidized bed combustion of rice husk to produce amorphous siliceous ash', *Energy for Sustainable Development*, International Energy Initiative, Inc., vol. 12, pp.33–42, 10.1016/S0973-0826(08)60417-2.

Rozainee, M., Ngo, S. P., Salema, A. A., Tan, K. G., Ariffin, M. & Zainura, Z. N. 2008b, 'Effect of fluidising velocity on the combustion of rice husk in a bench-scale fluidised bed combustor for the production of amorphous rice husk ash', *Bioresource Technology*, vol. 99, pp.703–713, 10.1016/j.biortech.2007.01.049.

Rubber and Plastics News. 2015, Goodyear reaches agreement for rice husk silica, June, https://www.rubbernews.com/article/20150610/NEWS/150619998/goodyear-reaches-agreement-for-rice-husk-silica.

Sánchez-Ramírez, E., Ramírez-Márquez, C., Quiroz-Ramírez, J. J., Contreras-Zarazúa, G., Segovia-Hernández, J. G. & Cervantes-Jauregui, J. A. 2018, 'Reactive distillation column design for tetraethoxysilane (TEOS) production: Economic and environmental aspects', *Industrial and Engineering Chemistry Research*, vol. 57, pp.5024–5034, 10.1021/acs.iecr.7b05058.

Schleiter, K. E. 2010, 'Silione breast implant litigation', *American Medical Association Journal of Ethics*, vol. 12, pp.389–394, 10.1001/virtualmentor.2010.12.8.medu1-1008.

Schneider, D., Wassersleben, S., Weiß, M., Denecke, R., Stark, A. & Enke, D. 2018, 'A generalized procedure for the production of high-grade, porous biogenic silica', *Waste and Biomass Valorization*, Springer Netherlands, vol. 0, pp.0, 10.1007/s12649-018-0415-6.
Shen, Y., Zhao, P. & Shao, Q. 2014, 'Porous silica and carbon derived materials from rice husk pyrolysis char', *Microporous and Mesoporous Materials*, Elsevier Inc., vol. 188, pp.46–76, 10.1016/j.micromeso.2014.01.005.
Silverman, G. & Rakita, P. 1996, *Handbook of grignard reagents*.
Siqueira, E. J., Yoshida, I. V. P., Pardini, L. C. & Schiavon, M. A. 2009, 'Preparation and characterization of ceramic composites derived from rice husk ash and polysiloxane', *Ceramics International*, vol. 35, pp.213–220, 10.1016/j.ceramint.2007.10.013.
Stevens, C. 1998, 'Environmental degradation pathways for the breakdown of polydimethylsiloxanes', *Journal of Inorganic Biochemistry*, vol. 69, pp.203–207, 10.1016/S0162-0134(97)10019-8.
Sujirote, K. & Leangsuwan, P. 2003, 'Silicon carbide formation from pretreated rice husks', *Journal of Materials Science*, vol. 38, pp.4739–4744, 10.1023/A:1027475018767.
Sun, L. & Gong, K. 2001, 'Silicon-based materials from rice husks and their applications', *Industrial and Engineering Chemistry Research*, vol. 40, pp.5861–5877, 10.1021/ie010284b.
Tang, Q. & Wang, T. 2005, 'Preparation of silica aerogel from rice hull ash by supercritical carbon dioxide drying', *Journal of Supercritical Fluids*, 10.1016/j.supflu.2004.12.003.
Tiwari, A. & Hihara, L. H. 2014, '*Sol-Gel route for the development of smart green conversion coatings for corrosion protection of metal alloys*', in Intelligent coatings for corrosion control, Elsevier Inc, 10.1016/B978-0-12-411467-8.00010-6.
Todkar, B. S., Deorukhkar, O. A. & Deshmukh, S. M. 2016, 'Extraction of silica from rice husk', *International Journal of Engineering Research and Development*.
Todkar, B. S., Deorukhkar, O. A. & Deshmukh, S. M. 2016, 'Extraction of silica from rice husk', *International Journal of Engineering Research and Development*, vol. 12, no. 3, pp.69–74.
Tour, J. M., John, J. A. & Stephens, E. B. 1992, 'Convenient route to di- and triorganosilyl ethyl ethers and the corresponding di- and triorganosilanes', *Journal of Organometallic Chemistry*, vol. 429, pp.301–310, 10.1016/0022-328X(92)83181-G.
Tullo, A. 2015, 'Firms glean the rice harvest for silica', *Chemical & Engineering News*, December, p. 19.
Della, V.P.[1], Kuhn, I. & Hotza, D. 2002, 'Rice husk ash as an alternate source for active silica production', *Materials Letters*, vol. 57, pp.818–821, 10.1016/S0167-577X(02)00879-0.
Varaprath, S. & Larson, P. S. 2002, 'Degradation of monophenylheptamethylcyclotetrasiloxane and 2,6-cis-diphenylhexamethylcycloterasiloxane in Londo soil', *Journal of Polymers and the Environment*, vol. 10, pp.119–131, 10.1023/A:1021146419479.
Wang, W., Martin, J. C., Zhang, N., Ma, C., Han, A. & Sun, L. 2011, 'Harvesting silica nanoparticles from rice husks', *Journal of Nanoparticle Research*, vol. 13, pp.6981–6990, 10.1007/s11051-011-0609-3.
Weiss, V. A., Reiff, G. & Weiss, A. 1961, 'Zur Kenntnis wasserbestandiger Kieselsaureester', *Zeitschrift fur Anorganische und Allgemeine Chemie*, vol. 311, pp.142–151.
Wong, D. P., Suriyaprabha, R., Yuvakumar, R., Rajendran, V., Chen, Y. T., Hwang, B. J., Chen, L. C. & Chen, K. H. 2014, 'Binder-free rice husk-based silicon-graphene composite as energy efficient Li-ion battery anodes', *Journal of Materials Chemistry A*, Royal Society of Chemistry, vol. 2, pp.13437–13441, 10.1039/c4ta00940a.
Wright, M. & Uddin, A. 2012, 'Organic-inorganic hybrid solar cells: A comparative review', *Solar Energy Materials and Solar Cells*, Elsevier, vol. 107, pp.87–111, 10.1016/j.solmat.2012.07.006.
Xu, S., Lehmann, R. G., Miller, J. R. & Chandra, G. 1998, 'Degradation of polydimethylsiloxanes (silicones) as influenced by clay minerals', *Environmental Science and Technology*, vol. 32, pp.1199–1206, 10.1021/es9708676.

Xue, Y., Wang, L., Shao, Y., Yan, J., Chen, X. & Lei, B. 2014, 'Facile and green fabrication of biomimetic gelatin-siloxane hybrid hydrogel with highly elastic properties for biomedical applications', *Chemical Engineering Journal*, 10.1016/j.cej.2014.04.049.

Yanagisawa, T., Yamaguchi, H. & Kobayashi, M. 1989, 'Absorbent of Chlorosilane Compound'. United States Patent Office.

Yi, E., Hyde, C. E. & Laine, R. M. 2003, 'Fumed silica from sustainable, green sources without first having to make SiCl4', *Manuscript in Preparation.*

Yilgor, E., Atilla, G. E., Ekin, A., Kurt, P. & Yilgor, I. 2003, 'Isopropyl alcohol: An unusual, powerful, "green" solvent for the preparation of silicone-urea copolymers with high urea contents', *Polymer*, vol. 44, pp.7787–7793, 10.1016/j.polymer.2003.10.048.

Zareei, S. A., Ameri, F., Dorostkar, F. & Ahmadi, M. 2017, 'Rice husk ash as a partial replacement of cement in high strength concrete containing micro silica: Evaluating durability and mechanical properties', *Case Studies in Construction Materials*, Elsevier, vol. 7, pp.73–81, 10.1016/j.cscm.2017.05.001.

Zavin, B. G. & Muzafarov, A. M. 2007, 'On the 100th anniversary of the birth of Kuz'ma Andrianovich Andrianov', *Polymer Science Series C*, vol. 49, pp.215–218, 10.1134/S1811238207030022.

Zhang, J., Chen, Y. & Brook, M. A. 2014, 'Reductive degradation of lignin and model compounds by hydrosilanes', *ACS Sustainable Chemistry and Engineering*, vol. 2, pp.1983–1991, 10.1021/sc500302j.

Zhang, J., Chen, Y., Sewell, P. & Brook, M. A. 2015a, 'Utilization of softwood lignin as both crosslinker and reinforcing agent in silicone elastomers', *Green Chemistry*, Royal Society of Chemistry, vol. 17, pp.1811–1819, 10.1039/c4gc02409e.

Zhang, J., Fleury, E. & Brook, M. A. 2015b, 'Foamed lignin-silicone bio-composites by extrusion and then compression molding', *Green Chemistry*, Royal Society of Chemistry, vol. 17, pp.4647–4656, 10.1039/c5gc01418b.

Roby Wong

7 Green possibilities in a green industrial sector: The lumber industry

Abstract: Industries have gone on far too long escaping the eyes of the general public on matters of sustainability and global environmental "greening". The lumber industry itself has gone on for hundreds of years without any indication of environmental consciousness, even with their main resource at a slow decline. From early human civilization to the present modern age, wood has been a longstanding staple as a natural resource and a symbol of environmental awareness. With the consumer public becoming more aware and conscious of worldwide concerns such as climate change, greenhouse gases, and other environmental issues, businesses are being condemned for their lack of initiative in preserving the global ecosystem for future generations. While the lumber industry is one such corporation, generations of wood technologies have contributed to the growing environmental effort of the business. From simply recycling wood scraps as fuel to processing wood composite boards for a myriad of different applications, these advancements to wood products and lumber processing have contributed exorbitant amounts of careful planning in order to construct a more sustainable, environmentally "green" industry.

Keywords: lumber industry, lumber production, industrial greening, green alternatives, wood-based products, waste utilization, wood waste, wood preservatives, life-cycle assessment, environmental assessment, natural degradation, physical properties (wood), mechanical properties (wood), sustainable use

7.1 Introduction

Durable, useful, and sustainable, the use of wood as a renewable source of raw material has been pivotal throughout human history. Wood has been the main driving power of industry and civilization, changing just as humans have changed, adapting just as humans have adapted. Due to its natural abundance worldwide, it has been harvested and used for a variety of products, from structural building materials to simple kitchen utensils. Because of its wide use throughout history and its recent resurgence as a renewable natural resource, it is imperative to assess the sustainability of the lumber industry. Recently, the demand for wood in large structural sizes and of high quality have increased as consumers become more environmentally conscious.

This article has previously been published in the journal Physical Sciences Reviews. Please cite as: Wong, R. Green possibilities in a green industrial sector: The lumber industry *Physical Sciences Reviews* [Online] 2020, 5. DOI: 10.1515/psr-2019-0025.

https://doi.org/10.1515/9783110597783-007

Wood is unique among all the different construction materials, modern, and traditional. Building materials have evolved from mud and stick huts to the colossal steel skyscrapers in cities. With modern materials such as steel and concrete, wood has a widespread appeal for a multitude of structural and non-structural applications that are favored more than modern materials such as steel, aluminum, and concrete. In addition, wood has numerous technical advantages over other construction materials: high strength-to-weight ratio, ease of jointing, ease of machining and handling, and many more [1].

Consumer demand has grown disproportionately large compared to the amount of timber available, as available timber from managed forests or plantations tend to be species that grow faster and less dense, hence a decrease in overall quality and in variability. Within the past 30 years, the global forest area has decreased from approximately 31% in the 1990s to approximately 30% by 2020 [2]. Although a difference of 1% over the globe does not seem significant, the amount of deforestation within this period has become a threat to the biodiversity of the world's forest and woodland, further solidifying the need for environmental precautions and "green" lumber conditions. Global forest areas include: areas with bamboo and palms; forest roads, firebreaks and other small open areas; forests in national parks, nature reserves, or other protected areas with specific scientific, historical, cultural, or spiritual interest; windbreaks, shelterbelts, and corridors of trees with an area of more than 0.5 hectares and width of more than 20 meters; plantation primarily used for forestry or protective purposes. The data shown in Figure 7.1 exclude tree stands in agricultural production systems, such as fruit plantations and agroforestry systems, urban parks, and gardens [2].

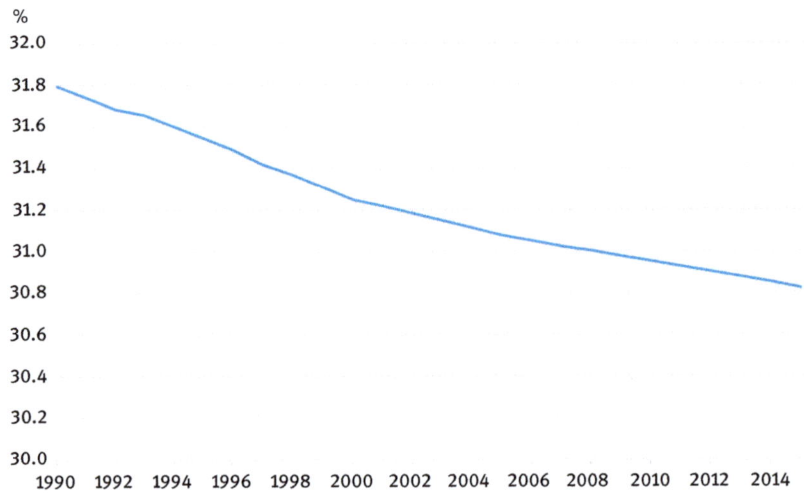

Figure 7.1: Timeline showing the steady decrease in percent of global forest area from 1990 to 2015.

There are many factors attributed to the decline of forested areas globally, such as wildfires, natural disasters, and disease, with human involvement the most destructive to woodland area. Deforestation, logging, and the lack of replanting harvested forest area is a big problem that is threatening the world's woodlands and in turn, the ecosystem. Because consumers are becoming increasingly aware of the damage older industries have inflicted on the environment, world business models are incrementally incorporating the concept of social responsibility and of sustainable development to their businesses, specifically the wood and lumber industries [3]. The growing concern about climate change has revitalized the interest in wood-based materials due to their low-carbon emission during production and ability to sequester carbon and allow forest carbon stocks to increase, thus providing an opportune moment for the wood and lumber industry to rebrand themselves as a globally sustainable industry.

In order to ascertain the economic feasibility of the lumber industry, life-cycle analyses (LCAs) can be run to determine whether or not lumber production is sustainable and cost-efficient. LCA is a broad examination of the environmental economic effects of a product at every stage of its existence, from harvesting to disposal, and beyond [4]. As indicated by recent LCA conducted over various species of wood, the overall mass retention of logs that go through the lumber refining process to produce planed, dry, hardwood lumber is approximately 45.8% of its original mass, while softwood lumber retains about 42.1% of its original mass [4, 5]. On account of consumer produced lumber being made without using the other 55% of the log, it is essential to explore options that allow for the entirety of the raw log to be processed into consumer-grade lumber with the minimal amount of waste possible.

7.2 How does the industry work?

Production of consumer lumber starts with the harvesting of logs from privately-managed forested areas or from tree plantations, then trucked to a lumber sawmill, scaled, graded, and stored in the log yard until they are ready to be sawn down to size. Logs may be stored wet or dry depending on species and season. Typically, that would mean the wood has to retain either less than 15% humidity or above 72% humidity to prevent fungal bodies, mold, or bacteria from altering the raw wood [6]. Figure 7.2 provides a rough overview of the general practice of lumber preparation and manufacturing within the lumber industry. There are four main processes in producing lumber: wood identification, sawing, drying, and planing, with energy generation as a "green" option throughout production to provide a sustainable practice.

Wood identification. The identification of wood is imperative for assigning quality markers on logs by recognizing the characteristic cell patterns and wood features for various different species of wood. Often logs from the same genus but different species show different properties and characteristics from each other, thus

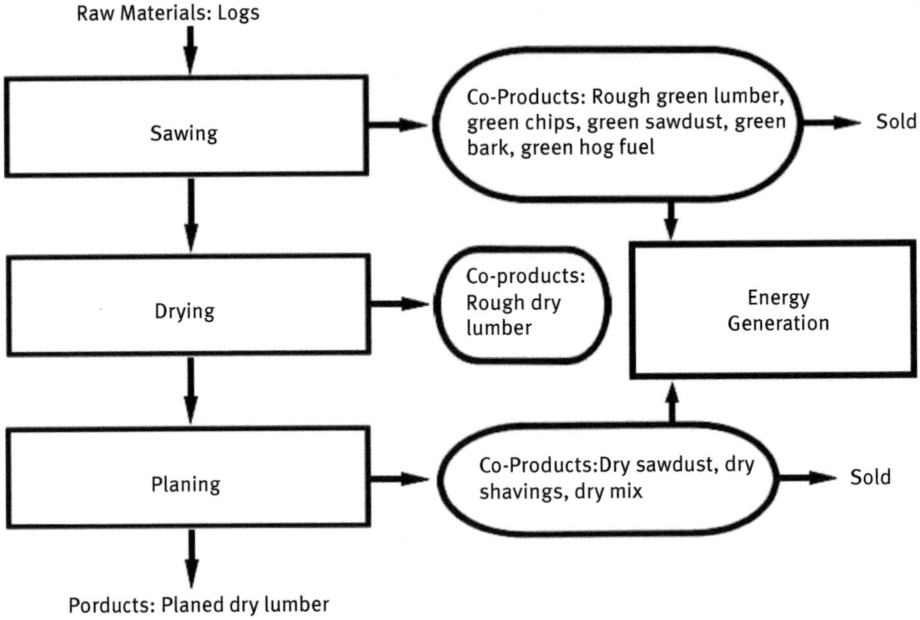

Figure 7.2: Depiction of lumber production from logs to end product [4, 5].

performing differently under various conditions and applications [6]. Because of this, serious problems can develop if species or genera are mixed during the manufacturing process and in use. Features often used to distinguish species include color, odor, grain patterns, density, and hardness [7].

The original method of grading wood was through the human eye, a.k.a. "visual" test. As this method is purely based on the outward appearance of the log, it is not accurate as defects on the surface are not indicative of possible defects in the structural properties that may persist as consumer-grade lumber. In addition, visual analysis with the human eye is not consistent with each person and each mill, leading to graded lumber that may be significantly above or below the actual quality of the lumber [8]. To supersede the shortcomings of the "visual" test, new technologies such as photo-electric scanning have been developed to not only accelerate the identification process, but to also provide more accurate and consistent results. This method involves moving logs beneath an optical scanning system that measures the characteristics of light reflected from the surface of the lumber. From there, it translates the refracted light into digital signals that are then analyzed to denote specific characteristics on the surface, i.e., knots, wane, cracks in wood [8]. The downside of photoelectric scanning as a method of wood identification is that it only analyzes the surface of the wood without relating to the structural properties of the lumber.

At this point, being able to identify and understand the structural properties of lumber in a quick and orderly fashion is of most importance. Therefore, it is necessary to know how much weight a piece of lumber can carry before it ruptures. In order to ascertain this information, an accurate analysis of the strength characteristics of the board is necessary. The point at which a structural product breaks is called its modulus of rupture (MOR) [9]. Similarly, to understand the properties of a specific piece of lumber, it is important to address a board's ability to withstand bending without breaking. The force required to deflect a board by a predetermined amount is called the modulus of elasticity (MOE) [9]. It is known that there is a direct proportionate relationship between a product's ability to bend and the point in which it breaks [9]. By analyzing a piece of lumber through its MOE and MOR, it is possible to combine the non-destructive test with the visual photo-electric screening to ascertain the true quality value of the wood, and prevent lumber from being graded incorrectly. The MOE test can be designed to provide digital output in conjunction with a predesigned program to optimize the cutting pattern of the lumber, therefore providing the optimum quantity and quality from the raw log. Moreover, the program can optimize cutting from large scale lumber to small scale, allowing the majority of the wood to be used and not wasted [8]. The use of these two tests in combination allows the majority of the wood to be identified for defects and cut accordingly to use the entirety of the log, showing a highly accurate approximation of the structural characteristics of the lumber and thus preventing excessive felling of trees. By measuring the MOE along the length of a piece of lumber to determine its structural characteristics and capabilities, scanning the surface of the lumber piece for visual markings to determine possible structural defects, and cutting the lumber piece into graded pieces as dictated by measurement data from MOE, MOR, and surface scanning, we are able to achieve the optimum number of pieces of predetermined value [8].

Sawing. Once the screening and grading process has been completed, logs will then be sawn to produce rough green lumber and wood residue accumulated from the sawing process. These include bark remnants, sawdust, wood chips, and edgings that alter the structural characteristics of wood. The majority of wood residue such as wood chips are sold as a coproduct to be pulverized and processed into paper, meanwhile other residual wood, primarily sawdust, is combusted on-site as fuel to kiln dry the lumber. The remaining wood residue can be sold off as compostable mulch, feedstock for particle board, or for other consumer goods [5].

Drying. After logs are sawn and stripped of their bark and observable defects, they go through a series of drying processes before finishing. This process begins with previously sawn, rough green lumber and ends with rough dry lumber going to the planer mill. Most lumber mills have their own specific drying process, whether it be air-drying in a shed, kiln-drying, or some combination of both. Drying time and methods can also vary depending on hardwood logs vs. softwood logs. Due to their high density and ability to retain more water, hardwoods such as red oak require longer drying periods than softwoods in order to prevent structural decay [10]. These variations in drying methods may affect the time it takes to dry lumber, the total energy consumption of the process, and whether or not the apparatus can recycle residual wood coproducts for energy. For some mills, it is possible for them to produce their own electrical power from wood residue produced on-site, consuming about four-times the amount of wood residue at mills that do not produce their own electrical power by recycling [4, 5]. By introducing wood scraps and other combustible materials derived from the sawing and planing processes, lumber mills would be able to exploit alternatives to petroleum and coal, reducing the carbon emissions from non-renewable fossil fuels to sustainable wood resources.

Planing. As the final stage of lumber production, it begins with rough, kiln-dried lumber and produces smooth, surfaced, packaged lumber sorted by specie type, lumber size, and the quality grade of the product. In addition, all the residual wood coproducts such as planar shavings, sawdust, and end trimmings are collected and repurposed as energy or sold as consumer products.

7.3 Environmental impact

One of the greatest attributes of wood is that it is a renewable resource. If sustainable forest management and harvesting practices are followed, this natural resource will be available indefinitely. For that reason, wood materials are environmentally favorable in place of other metal or petroleum-based materials. By placing an emphasis on wood products that are more economically feasible and environmentally friendly, increased regulation in the lumber industry would allow forests to thrive once again. Wood products produced from trees can range from a minimally processed log at a long-home building site to a highly processed and highly engineered wood composite manufactured in a large production facility [11]. On account of the variability in uses for wood materials, there are also environmental reasons to harness wood as an alternative in place of other widely-used materials.

Embodied energy. Embodied energy refers to the quantity of energy required to harvest, mine, manufacture, and transport the finished product [11]. Unlike metals and fossil-fuel-based materials such as plastics and other polymers, wood, a substance that requires a minimal amount of energy-based processing, has a low level of embodied energy relative to many other materials used in construction. Over half the energy consumed in manufacturing wood products in the US is from biomass or bioenergy and is typically produced from tree bark, sawdust, and by-products of pulping in papermaking processes [11].

Carbon impact. The role of carbon emissions in global climate change and predicted negative impact on ecological sustainability, as well as the general health of the planet have never been more aware to the general public. Forests play a major role in the Earth's carbon cycle as they are the net contributors of carbon to the atmosphere. Unfortunately, deforestation in woodland areas around the world is directly responsible for the release of stored carbon. Tropical deforestation is responsible for an estimated 20% of total human-caused carbon dioxide emissions each year, granted that is no excuse for the carbon dioxide emissions created through human processes, such as burning coal and fossil fuels [2]. Forests are sources of carbon when disturbed as a consequence of fires, insect damage, other reasons for natural mortality, or harvesting. They are sinks of carbon when recovering and regrowing after such disturbances [12]. Carbon in wood remains stored until the wood deteriorates or is burned and reduced to ash. To clarify, when a tree that remains in the forest dies, a portion of its carbon releases back into the atmosphere as the woody material decomposes. But when wood products near the end of its utility life, large structural building lumber for example, pieces can be recovered and reworked in another way, either repurposed into a new building material, chipped down for fuel or mulch, or sent to a landfill [11]. Carbon is transferred from forests to products when wood is harvested and used for various purposes. Otherwise, when using lumber to produce wood products or paper products, these commodities would store carbon while in use.

Wood substitution. To mitigate the increasing rate of carbon emissions that come from production industries such as steel, concrete, or other petroleum-based products, environmentally conscious companies have explored using wood as an alternative. Wood substitution implies any use of wood that replaces materials providing equivalent service or function to the original [11]. Therefore, increasing the transformation of forest biomass into wood products to replace products emitting more greenhouse gases per functional unit [12]. In particular, replacing petroleum-based products for wood products, or using wood in place of non-wood materials that are capable of influencing carbon emissions into the atmosphere. Carbon fluxes in the forest sector can be influenced directly by carbon stock changes in forests or forest products, and by substituting bioenergy for fossil fuels [12].

One particular industry that has the potential for substituting wood for other materials is the building and construction industry. Over the past decade, the concept of "green building" has become more mainstream due to public awareness at the potential environmental benefits of wood substitution in construction as opposed to conventional construction substances. Green building is defined as the practice of increasing the efficiency with which buildings use resources while also reducing the impact on human health and the environment. This can be achieved through better siting, design, material selection, construction, operation, maintenance, and removal throughout the complete building life cycle [11]. Furthermore, green building constitutes new technologies and methods that envelope the concept of environmentally sustainable construction. For this purpose, substitution between wood and competing building materials is of considerable interest.

Choosing building materials that show positive environmental attributes is a major area of interest to industries and consumer public because of the recent environmental movement. Wood has a plethora of positive characteristics, including low embodied energy, low carbon impact, and indefinite sustainability if regulated accordingly. These characteristics are important because in the US, a little more than half the wood harvested in the forest ends up as building material used in construction, while the other half are processed into miscellaneous wood products [11]. Wood has two significant advantages over non-wood substitutes; wood is carbon-neutral, and carbon can be sequestered through wood and wood products [4, 5]. Therefore, using wood as a fuel or in a finished wood product could be considered a sustainable practice. Although, to remain a sustainable and renewable practice, it is essential to modify current regulations to specifically focus on restoring wood resources strategically so that felling practices are not disproportionate to replenishing them. Presently, timberland and forested areas globally are at a decline with the regulations and practices that exist now. Gustavsson surmised that if current harvesting levels are maintained, forests will grow into older age groups and net increment of wood material will decline. Conversely, if harvesting levels are increased, age class structures will change towards younger age classes and therefore net increment will increase [12].

Including practices such as optimization of thinning and harvesting operations, fertilization of smaller plants, choice of tree species, or increasing the forest surface area would also suffice in maintaining the current rate of harvesting. The main problem hindering this is the consumer supply and demand of some species of wood over others, as well as preference over specific types of species for different applications. Hardwood species are of major concern for sustainability as the consumer demand for hardwoods are disproportionately high compared to the amount of hardwood that is harvested. Hardwoods typically take a longer time to grow, consequently prolonging the turnaround time of planting and harvesting hardwood species. Specific species of trees, where it takes a much shorter time from seedling to harvest, are then used to replace hardwoods that have been felled without being able to realize the economic demand for hardwood species. For that reason, researchers have spent the past decade exploring and exhausting different ways to incorporate sustainability into the lumber industry through in-depth study of wood degradation and the new technologies to preserve the lumber and wood products that already exist.

7.4 Decay and degradation

All wood is composed of various ratios of cellulose, lignin, hemicellulose, and small amounts of extraneous material within a cellular structure [10]. Variations in the characteristics and proportions of these components in the cellular structure contribute to the physical and structural characteristics that separate species from one another. These variations can affect the density, their MOR and MOE, or whether they are hardwoods or softwoods [10]. The two different types of wood differ not only in the tree species they are derived from, but also their cellular components.

Softwoods have a simpler basic structure than hardwoods because they only have two cell types, fibrous tracheids cells and parenchyma ray cells, with little variation in structure [7]. Hardwoods have greater structural complexity due to an increased amount of basic cell types, consisting of vessels, fibres, and parenchyma, which allows hardwoods a higher degree of variability within the cell [7]. The single most important distinction between the two general kinds of wood is that hardwoods have a characteristic type of cell called a *vessel element* (or pore) whereas softwoods lack these, as seen in Figure 7.3. Although the quantity of cellulose remains quite consistent at 40–50% cellulose in both hardwoods and 40–45% cellulose softwoods, hemicellulose ratios vary the most [15]. Hemicellulose in softwoods range between 25 and 30%, while hardwoods contain 25–35% hemicellulose [15]. Normally hardwoods have a lower lignin content of 20–25% lignin as compared to softwoods with 25–35% lignin [15].

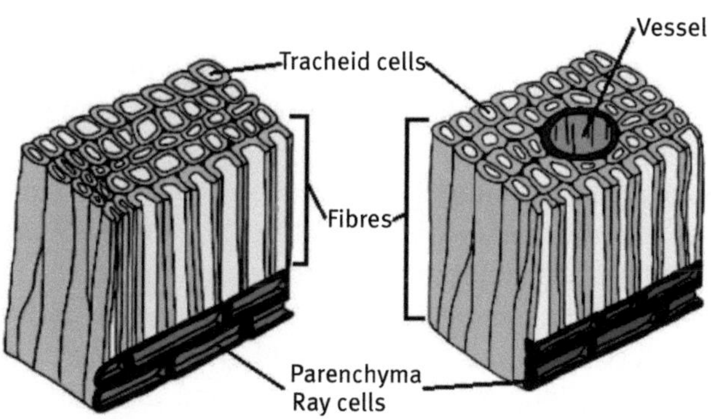

Figure 7.3: Diagram of softwood cells (left) and hardwood cells (right) [13, 14].

Lignin plays a governing role in natural wood protection against wood-destroying organisms (WDOs), with high lignin levels considered to provide better natural protection than low lignin levels. Of this, guaiacyl lignin and syringyl lignin, illustrated in Figure 7.4, are the main structural lignin when considering a specific species' ability in preventing cell degradation [15]. Guaiacyl lignin is composed of coniferyl alcohol units, predominately found in softwoods and are typically harder to digest, while hardwoods contain some ratio of guaiacyl lignin and syringyl lignin [16]. Syringyl lignin is composed of sinapyl alcohol units, making it easier to pulp [16]. Because the distribution of syringyl and guaiacyl lignin can vary in different cell types within hardwoods, the ratio between the two heavily influences the manner and speed at which the plant cells and tissues are degraded by WDOs.

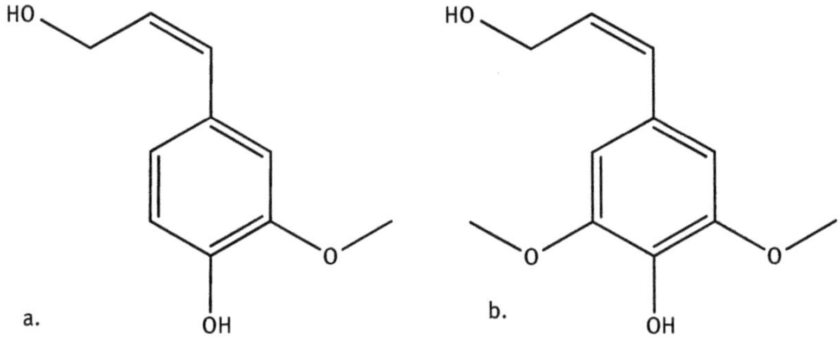

Figure 7.4: Molecular structure of the main components in (a) coniferyl alcohol in guaiacyl and (b) sinapyl alcohol in syringyl lignin [16].

The individual wood cells may also vary in degree and type of lignification, for instance, the fibers in hardwoods are syringyl lignified while vessel walls tend to possess guaiacyl lignin, providing dual protection against WDO. Both the variability and ratio of different cell types as well as presence and type of lignin in hardwoods, and to a lesser extent softwoods, play a major role in regulating biodegradation [17].

The biological degradation of wood is a natural process of organic matter recycling and occurs in all natural environments, provided adequate conditions are available [17]. Oftentimes, the extent of wood degradation is directly proportional to the exposure of the substance to various natural elements; water presence, moisture, excessive heat, salt, sunlight, temperature variation, mold, bacteria, fungi, or insects. Specifically, temperature variation, moisture or wetting, and rapid drying are major causes of wood decay within the lumber industry, as they cause wood to swell and shrink [18]. Moreover, WDO such as fungi, mold, or bacteria commonly inflict damage to the surfaces of raw logs and lumber alike causing damage to the structural integrity of the wood. Under natural conditions, wood may be colonized by an assortment of these decay-inducing microorganisms [17]. These WDO initiate the process of biodegradation where the wood structural components are solubilized until all that is left are carbon dioxide and water [17]. Biodegradation of wood may result in physical deterioration, color change, surface roughening, and eventually internal damage to the wood microstructure [18].

That being said, each species reacts and degrades differently to each degradation factor, just as the manner in which wood is degraded can vary according to the WDO involved, species of wood, nature of the environment, as well as whether the WDO are competing with one another [17]. It has been shown that environmental degradation and biodegradation can produce significant changes in the mechanical and morphological properties of wood. Below are various examples of both natural and biological degradation that are apparent within the lumber industry:

Air, sunlight, and temperature degradation. Research has indicated that the surface of wood can be altered in conditions where the wood is repeatedly wet and dry, is exposed to high and low temperatures, and to direct sunlight. In places with high humidity, the water in the air gets absorbed by the wood surface, creating a moist environment in the wood. When dryer air blows through, it removes moisture on the surface at a rapid rate, causing high fluctuations in swelling and shrinking of cells. Extreme moisture and temperature conditions induce degradation of the wood surface through expansion and contraction of the cells, which in turn causes roughening of the wood surface, checking, splitting, and wood cell erosion [18].

Earth and water degradation. Earth degradation is representative of a combination of effects caused to wood by air, water, and sun. Under direct sunlight, the earth dries and due to reverse osmosis, it causes the wood buried in it to also lose water at a rapid rate, causing extreme stress on the cells. Under wet or rainy conditions, earth will concentrate water around the wood and keep the wood wet for a much longer period of time than if it were outside of earth. This phenomenon will allow water to get through the wood via osmosis and saturate it, causing dimensional instability and damage to fibers. In addition, water is one of the necessary mainstays of microorganisms, such as bacteria or decay fungi, to breed. Therefore, the presence of water would provide a more habitable surface for decay fungi to take root in [18].

Fungal degradation. Wood is very prone to biodegradation by an assortment of microorganisms such as molds, bacteria, and primarily fungi. These microorganisms feed on wood for nutrition within the cells, degrading and altering the mechanical properties of lumber and wood products [18]. As seen in Figure 7.5, the multiple layers of wood oftentimes provide surface protection against environmental degradation. However, these microorganisms have each evolved their own method of penetration and degradation of these barriers. Fungi generally play a dominant role in wood decay, though there are certain groups of bacteria that are able to degrade the lignin within cell walls throughout the woody substance. In environmental conditions with adequate moisture and oxygen, both fungal and bacterial attacks are present. For decades, identification between fungal degradation and bacterial degradation was nearly impossible. Recent advances in biological technology illuminated specific degradation patterns produced by bacteria that are strikingly dissimilar from decay patterns formed by fungi [19]. While most WDOs colonize cut wood surfaces, there is a tendency for all decay species to colonize the rays of the wood first, and then grow further into the wood structure via natural openings or directly from cell to cell by producing bore holes through the cell walls [17]. As aforementioned, lignin is a major component within the cell walls of wood and an important factor against biodegradation. Because of this, WDOs have developed methods for enzymatic degradation of lignin, despite lignin's chemical properties that prevent most decay microorganisms from destroying the cell walls. Between fungal and bacterial microorganisms, fungi with the ability to negate the effects of lignin are more common than the latter.

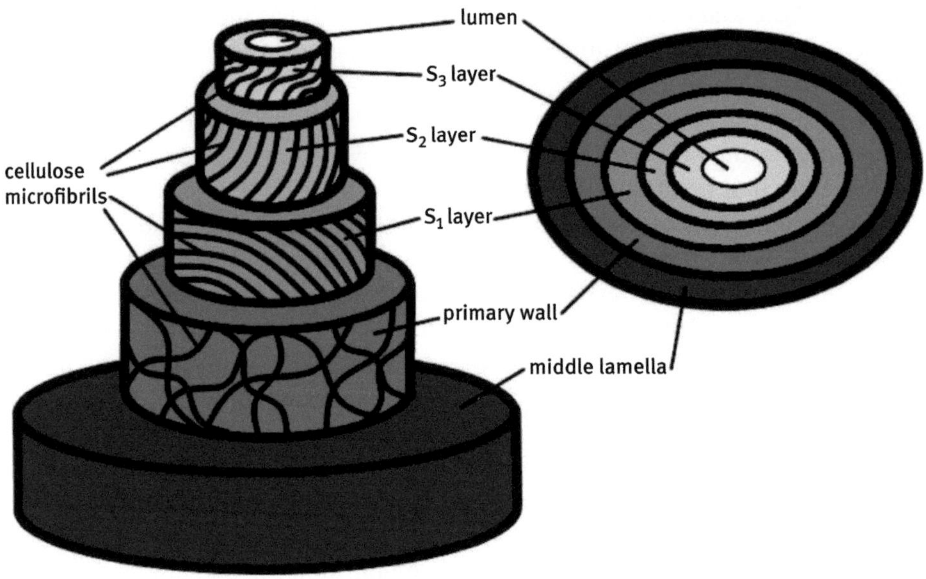

Figure 7.5: Diagram of cell wall layers of wood cells [20].

Soft rot fungi. The term "soft rot" is used to describe a large amount of fungi that typically produce chains of cavities within the cell wall layers of both soft and hardwoods, causing degradation patterns that roughen up the surface. Rot fungi in particular use specific mechanisms to overcome the lignin barrier of wood cells. Soft rot fungi align themselves along the cellulose microfibrils of the cell wall and overcome the lignin barrier by maintaining an extremely close position adjacent to both cellulose and lignin, allowing cellulolytic enzymes of soft rot fungi to be within proximity to the wood cell wall. Soft rot erosion decay causes a characteristic thinning of wood fiber walls from the cell lumen simultaneously with cavity formation [17]. Soft rot fungi typically cause a degree more decay in hardwood than in softwood, which is thought to be related to the higher lignin content and guaiacyl type of lignin in softwoods, compared with the lower and syringyl-guaiacyl nature of hardwood lignin [17].

White rot fungi. White rot fungi has been spotted attacking predominately hardwood species in both terrestrial and aquatic environments. Colonization by white rot fungi occurs via the rays, then penetration between cell walls by bore holes. Because white rot fungi are not restricted by the lignin barrier and are able to degrade all the wood components progressively from the cell lumen, it can cause major damage to the structural integrity of the wood substrate [17]. There are two main morphological decay types that white rot fungi use, described as "simultaneous" and "preferential". White rot fungi may preferentially attack the lignin and hemicelluloses to bore holes, or degrade all the wood components simultaneously from the cell lumen outwards, slowly progressing into adjacent cell walls [17].

Brown rot fungi. Alternatively, brown rot fungi have developed their own mechanism in bypassing the lignin barrier in cell walls. Brown rot fungal attack is characterized by excessive degradation and depolymerization of cellulose and hemicelluloses, leaving behind a weak, residual skeleton that eventually crumbles when dry [17]. When the shell of wood cells degrade, the damage to the lumber is instantaneous. For this reason, brown rot fungi is considered the most significant microorganism of all WDOs, as the presence of the fungi will immediately affect the structural quality of lumber.

Bacterial degradation. Certain groups of bacteria have evolved mechanisms and adaptations to also degrade lignified-cell walls in wood and produce degradation patterns that serve as diagnostic features for the identification of bacteria-caused cell wall degradation [19]. Generally, the bacteria degrades hemicellulose first, followed by cellulose, hardwood lignin, and then softwood lignin. Various wood products around the world have been found with signs of bacterial degradation, regardless of specie of wood or environmental conditions. When compared to rot fungi, wood degrading bacteria seem to possess a much higher tolerance to toxic chemicals. Moreover, these same wood degrading bacteria have been found colonizing and successfully degrading some preservative-treated wood. These bacteria have been identified and named according to their main method of degradation; tunneling bacteria (TB) and erosion bacteria (EB). Both TB and EB can attack a variety of substrates, with distinct differences in their form and mode of attack [19].

TB. TB are named as such due to tunnels formed as they degrade the cell wall. By breaking down lignified cell walls through tunneling, they are able to fully surround themselves by the substrate. This specific method of degradation enables TB to stay in close proximity to cell wall polymers through the use of extracellular slime surrounding the body of the bacterium [19]. The thin layer of slime allows the bacterium to glide through the tunnels they produce. After multiplying in the cell lumen, TB are able to spread out and expand the attack in the wood substrate. TB tend to have rounded ends, providing maximize degradation while conserving their energy, as well as uptake any leftover elements of the cell wall (i.e., enzymes and radicals) [19]. The tunneling mechanism also allows TB to explore all available cell wall areas and locate optimal areas of engagement. Doing so facilitates cohabitation between TB and other wood degrading microorganisms, such as EB or soft rot fungi. Moreover, TB have been observed to survive and thrive in wood substrates typically resistant to white and brown rot fungi – a strong indication that TB not only degrades lignin in cell walls, but also degrades recalcitrant substrates typically resistant to microbial attack [19].

EB. EB are named accordingly for their degradation method by eroding the exposed faces of the cell wall, producing hollow erosion channels. These erosion channels are parallel with cell wall microfibrils, allowing the bacterium to sink into the erosion sites while degradation occurs [19]. There, EB can colonize the cell lumen, multiply, and broaden the attack regardless of lignin composition, although EB are unable to degrade the lignin-rich middle lamella in normal wood and the highly lignified outer regions of compression wood cell walls [19].

Similar to TB, EB are able to excrete an extracellular slime coating to keep the bacterium adjacent to the cell wall and ensures that the bacteria cannot be sequestered in moist or wet environments. By surrounding itself in a slime coat, cell wall residual agents can diffuse through the slime sheath and to the bacteria in the degradation site [19]. But unlike TB, EB possess flat to angular or conical ends that help with erosion. EB have been observed to coexist with TB on chemically protected wood products, indicating that EB can also tolerate recalcitrant substrates used to treat wood, although EB are able to degrade wood while in oxygen-deprived conditions whereas TB are only restricted to oxygen-rich environments.

Of minor concern, mold and stain fungi are present in addition to the WDOs mentioned before, but they do not cause structural damage to lumber or wood products. Mold and stain fungi can grow on the surface of wood independent from other microorganisms, if freshly sawn wood is stored under high humidity and proper temperature. Most of the attacking mold and stain fungi produce pigmented spores or hyphae that discolor the surface or interior of wood [21]. This discoloration is due to coloring compounds produced by the reaction of the wood components with the secretions of the microorganisms. Though discoloration is the main concern with these staining microorganisms instead of debilitating structural effects, mold and staining fungi can affect the lumber industry. Molds found in homes, hotels, and school buildings may cause chronic sinus infections, respiratory infections, asthma, inflammation, and injury in gastrointestinal and pulmonary tissues [21], while discoloration of lumber may avert consumers from purchasing effected, but otherwise structurally sound wood products.

Unprotected wood in natural environments are rapidly degraded by WDOs such as fungi and bacteria, altering the structural and physical integrity of wood products. Unfortunately, the essential properties of wood can also be deteriorated by various environmental factors, which is the main disadvantage for wood and heavily limits its applications. By understanding how microbial degradation happens to wood and how it affects the structural integrity, protective strategies against WDOs can be developed to enhance and elongate the service life of wood products.

7.5 Coatings and preservatives

Preservatives are the mainstay treatments that extend the life of wood products, which helps to subsidize the amount of lumber harvested from forests. In using preservatives, lumber lasts longer and prevents microbial degradation of wood products. Preservatives can prolong the shelf-life and service-life of lumber, therefore reducing the excessive harvesting of forest resources. By preventing lumber from rotting away in storage, in residential homes, and in the outside environment, drawing out the utility-life of wood products help with sustaining the industry as well as timberland. The application of chemicals is an effective way to incorporate pesticides, antifungal, antimold, and antibacterial agents into wood to prevent the destruction of structural properties and to prevent discoloration on the surface. But with such an array of formulation possibilities, it is imperative to select chemicals based on "green" properties such as low toxicity, low solubility, and non-volatile compounds. Decades of preventative formulations have been used to sedate the degradation of lumber and wood products, whether or not they are entirely benign to the environment.

The main types of wood preservatives that are used to act as fungicides and pesticides for lumber and other wood products are either oil-borne or water-borne preservatives [22]. The effectiveness of a specific treatment depends on the formulation selected, method of application, proportion of sapwood to heartwood, moisture content, amount of preservative retained, depth of chemical penetration, and distribution of chemical within the wood [22].

Oil-borne preservatives contain chemicals that are usually insoluble in water, and are therefore dissolved in either petroleum or other organic solvents to facilitate wood penetration. The major problem with using oil-borne preservative treatments is specifically due to the need to dissolve chemicals in solvents such as petroleum. The use of petroleum and other organic solvents raises environmental issues on the production, use, and disposal of oil-borne preservatives. To mitigate possible biohazards associated with petroleum, recent research developments have made available oil-borne preservatives formulated as water in oil emulsions or dispersions in water [22]. Despite possible environmental issues, oil-borne preservatives have been proven toxic to fungi, insects, and mold, acting as a strong preservative treatment against WDOs. Because these treatments are dissolved in oils, it also provides a wider range in viscosity, vapor pressure, and color, allowing ease of control when formulating the preservatives [22]. Although due to the oily nature of petroleum, wood products treated with oil-based preservatives retain an oily, unpaintable surface that emits strong toxic odors irritating to plants, animals, and humans alike. Oil-borne preservatives also have a tendency to bleed or exude substances from the wood surface, which may lead to environmental biohazards to the surrounding areas the product comes in contact with [22]. Most oil-based preservative treatments have been cycled out and replaced with newer water-borne preservatives, though a few still exist due to their phenomenal antimicrobial properties.

On the other hand, water-borne preservatives are mainly composed of compounds such as copper, chromium, arsenic, and fluoride suspended in water. Several formulations involving combinations of copper, chromium, and arsenic have shown high resistance to leaching, as well as effective performance as a preservative treatment [22]. Water-borne preservatives are also competent as a fire retardant and typically contain compounds that have low toxicity towards humans, making it a safe treatment for interior applications. Although when used as a treatment for exterior wood products, there have been instances where leaching may occur due to excessive weathering [22].

Regardless of how preservative treatments are synthesized, decades of formulations have produced some chemical treatments that are particularly detrimental to the environment. The major issues with these preservatives are the extreme health concerns that come with them, such as a specific chemical in their formulation that are hazardous to the environment to organisms.

Creosote, for example, is an oil-borne preservative that has been used since the 1940s as a heavy duty wood preservative. The primary compound is distilled from coal tar at immensely high temperatures, producing a heavy, oily liquid [23]. Creosote has been proven to improve the weathering characteristics of wood and provide extremely concentrated protection from insects and fungi. It also promotes insolubility in water, providing adequate effectiveness in aquatic environments. Treatment products containing creosote as the active ingredient are used to protect wood against termites, mites, fungi, bacteria, mold, and any other organisms that may degrade or threaten the structural characteristics of lumber and wood products alike [23]. Although creosote is still used to treat railway ties, it is known that creosote shows a fair amount of bleeding, leaching, and evaporation from the wood. This poses a risk for both the people that handle the formulation and execution of creosote, but also to the consumers that may come in contact with the preservative. For that reason, the EPA has dictated that creosote is a Restricted Use Product (RUP) – for commercial purposes only [23].

Another example of a particularly hazardous preservative treatment developed is pentachlorphenol (penta). A crystalline compound dissolved in light petroleum oil, penta was one of the most widely used biocides and wood preservative in the United States. After it was found that penta is extremely toxic to humans, it is now classified as an RUP and no longer available to the general public [24]. Acute ingestion or inhalation of the compound resulted in neurological, blood, and liver effects, as well as an eye irritant to humans. Chronic exposure to penta have resulted in health issues to the respiratory tract, blood, kidney, liver, immune system, eyes, nose, and skin [24]. Not only were these health dilemmas found to affect those handling penta, but also those that have been exposed to wood products treated with penta preservative. As a results, the EPA now classifies penta as a Group B2 compound, probable human carcinogen.

Within the past few years, researchers have developed other wood preservatives that are much less toxic to humans and the environment. These preservatives offer alternatives to the previous options above, although they possess their own problems. Basically, a number of these contain chemicals or chemical byproducts that may cause detriment to humans and the environment if not used by its intended use.

Chromated copper arsenate (CCA) is a water-borne chemical preservative that has expanded into primarily three standardized formulations: CCA Type A, CCA Type B, and CCA Type C (CCA-C). CCA is generally used to treat wood used aboveground, in contact with the ground, or in contact with both freshwater or seawater. The effectiveness of CCA is apparent, in service from the late 1970s until 2004. Even now, CCA is regarded as the reference preservative used to evaluate the performance of other water-borne wood preservatives [25]. CCA-C is the most widely-used formulation, owning to is resistance to leaching and observable effectiveness at protecting against WDOs. CCA-C is composed of 47.5% chromium trioxide, 18.5% copper oxide, and 34.0% arsenic pentoxide dissolved in water [25]. Despite decades of proven performance, CCA has been phased out voluntarily for most applications around residential areas where human contact is prevalent, as CCA contains inorganic arsenic [25]. In 2004, it was deemed that CCA be classified as a Restricted Use Product (RUP) by the EPA.

In contrast, alkaline copper quaternary (ACQ) compounds are one of several water-borne wood preservatives that have been developed recently as a replacement to CCA. Although many variations of ACQ compounds have been produced and standardized for a myriad of applications, the majority contain 67% copper oxide and some quaternary ammonium compound as the main components due to their anti-microbial and pesticide properties [25]. The different formulations of ACQ allow some flexibility in achieving compatibility with specific wood species and application: ACQ Type B (ACQ-B) is an ammoniacal copper formulation, ACQ Type D (ACQ-D) is an amine copper formulations, and ACQ Type C (ACQ-C) is a combined ammoniacal-amine formulation with a slightly different quaternary ammonium compound [25]. ACQ preservatives are widely used due to their low toxicity to humans, though the composition of the wood preservative has been proven to accelerate corrosion of metal fasteners for in-service wood products which may have some effect on the environment over an extended period of time [25].

The most recent and widely received preservative treatments are borates and quaternary ammonia compounds (QACs). These water-borne preservatives consist of compounds that are low toxicity to humans and are naturally derived in the environment. Though both preservative compounds are incapable as standalone preservatives, when used in conjunction with other formulations, they provide a much more extensive protective treatment against pests and microbial WDOs.

Borate compounds are the most commonly used, unfixed water-borne preservatives in the lumber industry. Countless construction materials nowadays are treated with borates for typically non-biodeterioration treatments such as fire retardancy and insect control; however, recent research has concluded that borates could also be used to reduce mold growth on lumber [26]. Boron has additional characteristics that make it a viable, green preservative in lieu of other alternatives, such as its natural activity against fungi and insects, but low mammalian toxicity. Inorganic boron is listed as a wood preservative in the American Wood Protection Association (AWPA) standards, which include formulations prepared from sodium octaborate, sodium tetraborate, sodium pentaborate, and boric acid, with the most common as disodium octaborate tetrahydrate (DOT) [26]. DOT has higher water solubility than other form of borate, allowing more concentrated solutions to be used and increasing the mobility of borate through the wood [25]. In order to fully use the potential of borates as a preservative treatment, it is essential for the wood to retain high rates of borate compound in the wood and on the surface. Borate formulations are typically used as surface treatment for a wide range of wood products, but the disadvantage of borates is that they are unfixed. Unfixed preservatives can leach from treated wood over time. Boron's ability to diffuse with water into wood makes it more preferential than other traditional preservative treatments, though the same ability allows boron to leach out of the substrate as well [25]. Fortunately, boron naturally occurs at relatively high levels within the environment, and even in higher concentrations, borates have very low toxicity towards humans, mammalians, birds, aquatic invertebrates, and fish [25].

Commonly used in conjunction with borates, QACs have long been used for their efficacy against WDOs in wood treatments. Additionally, QACs have been found to contain low mammalian toxicity whilst destroying microbial colonies and preventing bacterial and fungal degradation [27]. The two main QACs widely used in wood preservatives are dodecyl dimethyl ammonium chloride (DDAC) and dodecyl dimethyl ammonium tetrafluroborate (DBF). DDAC is currently used as a component of ACQ wood preservative to improve its fungicidal properties [21]. While DDAC itself is not enough to completely protect wood from most WDO, such as soft rot and bacteria, it has the potential to prevent mold and staining fungi from discoloring lumber and marring its surface. DBF on the other hand has been recently developed to include tetrafluroborate into its formulation as a preservative that would be able to protect wood against mold and staining fungi, as well as termites and other wood destroying insects. Both DDAC and DBF have been proven to provide protection against mold and stain fungi, and are capable at very low concentrations, and are capable in protecting wood used outdoors against fungal decay, mold, stain fungi, and termite attacks [21].

Moving from environmentally detrimental, hazardous preservatives towards "green" formulations for wood treatments is indicative that lumber production is beginning to establish itself as a sustainable, ecological industry. Despite the monumental advancements in wood preservative treatments, the decline of forested areas is still ongoing. The use of preservative treatments are beneficial to elongating the service life of wood products, but they are still not enough to sustain a truly self-efficient industry. After decades of focusing resources in preservative treatments, new technologies have been discovered.

7.6 Wood-polymer composites

Though preservatives have been beneficial for the past century in treating and preserving wood products for a myriad of applications, there has still been a significant push to look for alternatives other than wood. Alternatives that have the same structural properties as wood while still environmentally sustainable are few and far between. Materials such as aluminum, steel, and concrete have been used as replacements for wood, but the production of these building materials do not benefit the global environment, nor are they feasible as a self-sustaining resource. In order to provide the necessary resources needed for structural applications, it is imperative to source alternatives from reliable natural resources, such as wood. Recent developments within the past few decades have produced much more economical and "green" products to act in place of wood as structural components. The ability to use forest resources in a more beneficial product via reconstituted wood composite technology provides reassurance in reviving and providing this natural resource for generations to come.

Developed in the 1970s, laminated veneer lumber (LVL) is an all-veneer engineered wood product and now one of the most widely used alternatives to solid lumber products. In a market where high quality solid wood lumber has been progressively scarce and difficult to find, new building material technologies, such as LVL, are considered the new norm within the lumber industry. In an effort to produce an environmentally efficient wood product, LVL is considered to be superior to recycled steel and plastic, concrete, stone, and brick in terms of energy requirements to produce [28]. Veneer are extremely thin slices and slivers of wood that can be obtained from nearly any species of tree. Similarly to plywood, LVL is composed entirely of veneer oriented in the same direction for maximum strength and stiffness [29]. Due to its parallel-set veneer, it has also been called parallel-laminated veneer (PLV).

Most LVLs are made by laminated, rotary-peeled veneer, increasing the yield for each log by 25–50% due to the use of the entire log [28]. When compared to traditionally sawn lumber, a large percentage of the log is lost due to the squaring of round logs. Similarly to plywood, LVL are produced by peeling large logs to create large veneer sheets. These sheets are stacked to the desired thickness, coated with a waterproof adhesive, and then exposed to extreme heat and pressure to form the final configuration of laminated veneer panels. The veneer panels are then cut to produce standard sized LVL for a variety of applications [28]. Because of its ease of handling during the entire production process, LVL can be manufactured in longer desired lengths or greater widths and thicknesses while maintaining the uniformity from the veneer [28], taking advantage of its unique structural characteristics. Because veneer products are composed of multiple layers of veneer, natural defects common in traditional lumber such as knots, holes, cross-grain, cracks, and splits are dispersed throughout the entire veneer board. This provides much less variations in strength, enhancing the structural integrity of the board. Similar to traditional solid lumber, LVL also shrinks and swells when exposed to a range of weathering and temperature fluctuations. However, because of its uniform composition, when LVL shrinks or swells, the entire board will shrink and swell uniformly causing little to no complications with using LVL or most wood applications [28]. In addition, LVL are typically used in nonvisual industrial applications, allowing it to be concealed entirely or wrapped for a desirable outward appearance. Because of this, it is possible to produce LVL through any specie of wood, allowing the use of typically unfavored species for traditional sawn lumber.

There are multiple reasons believed to have ignited LVL research; the diminishing supply of large logs suitable for manufacturing lumber of large dimensions, the economic feasibility of using a larger percentage of the forest biomass when harvesting previously considered forest residue, or the cost advantage of complete use of all materials entering processing [28]. Regardless, production of LVL incentivized the use of smaller logs that would otherwise have been unused entirely. Other products similar to LVL include plywood, where the veneers switch direction while stacking into blocks.

In a similar vein to LVL, glued laminated panels (glulam) is a wood composite made with thin laminates that are glued together to form a material with much more uniform distribution of fibers and higher values of mechanical characteristics than wood [30]. In addition to high design stress value at the stake of lower variability, glulam can be bent during production to construct curved profiles of wood, as well as individual beams and columns for construction [1]. In some cases, using specific adhesives characterized by resistance to temperature, climate changes, chemicals, and microorganisms can provide protection to glulam products. Oftentimes, adhesive properties give glulam construction an advantage over traditional reinforced concrete and steel structures [30]. When compared to solid timber production, glulam has multiple advantages; for instance, pieces of large cross sections for glulam panels can be easily produced, laminations are quick and cheap to process, elongation of glulam panels are possible through end jointing, significant reduction in distortion and splitting of the wood, and wood defects in laminates are dispersed throughout the board [1]. Cross-laminated timber (CLT) is made similarly to glulam, with the exception that CLTs are typically made with smaller cuts of timber that are placed in alternating orientations to form an odd number of layers. Each layer is oriented perpendicular to adjacent layers and glued together under pressure, while glulam has all of the layers oriented parallel to one another.

Glulam are one of the lightest construction materials within the industry as they are rigid and durable, and are resistant to humidity, temperature fluctuations, and WDOs. Glulam consists of laminates, which are long timbers organized to form small cross-sections in the wood bonded together under pressure with structural adhesives [1]. The laminates are arranged parallel to one another and properly oriented according to their growth rings to increase its elastic and mechanical characteristics. To elongate glulam panels, laminates may be jointed together through the use of structurally sound joints like the wedged finger joint [30]. Multiple variations and formulations of glulam have been developed for a variety of applications. Generally, any type of wood can be used for glulam construction, so long as the wood is porous enough for the adhesives to penetrate the cells [30]. Depending on the specie, preparation of wood is an important aspect for wood adhesion, as it can cause significant changes in the wood characteristics and to the final glulam product. Wood drying is also an important factor that can affect the mechanical processing, wood stability, and in the possible creation of internal stresses after adhesive bonding [30].

Aside from being a versatile construction material, there are significant environmental advantages of glulam over other materials. The possibility of preventing wood biodegradation while being able to effectively recycle the composite material proves to be much more "green" than traditional building materials [30]. Conversely, the production of glulam components requires roughly twice the amount of energy to construct solid wood lumber. In addition, the adhesives necessary for the mechanistic and structural benefits of glulam contain formaldehyde, which is harmful for the human health and the environment [30]. However, adhesives account for approximately 1% of the entire volume of production, and thus not considered to be a problem. Future prospects for glulam include the optimization of its production and eventually use weaker wood or recycled wood for laminate production [30]. Ultimately, glulam currently fulfills a portion of the consumer demand for traditional wood products whilst making use of underutilized and undervalued trees.

While both LVL and glulam exploit the thin, young trees commonly grown in tree plantations, there are few solutions for handling all the excess wood residue not sold as a coproduct. Sawdust, bark remnants, wood chips; all conveniently produced during the lumber process but unfavorable for consumer products. That is until the introduction of particleboard technology, leading to the development of the oriented strand board (OSB). OSB is the culminating consumer product from generations of particleboard technological development. Starting in the 1920s, particleboard was initially made from sawdust and wood scraps adhered together through the use of adhesives. Now, OSB are three-layered flakeboard panel, cross-banded by a core of oriented wood strands (flakes) longer than 3 in. and approximately 0.0015 in. thick [31]. OSB production includes reception and storage of wood logs ("logging"); debarking and stranding of logs ("flaking"); drying of wood strands from 50–150% of moisture content to 5% ("drying"); screening of dry strands to separate the finest sizes ("screening"); blending of stands with adhesives mixture, wax, and hardener ("chemicals"); forming in order to obtain a mat "blending and forming"); pressing of the mat to get the OSB ("pressing"), followed by an optional final step of sanding and finishing the OSB panel [32]. The use of thin and long flakes allow overlapping with each other to provide structurally sound composite boards. OSB boards are typically built in three layers of flakes bonded with resin and oriented with perpendicular strands interwoven throughout the board [32]. By designing the board's flakes this way, the panels are structurally sound with strong mechanical properties for application in multiple industries, such as building, packing, and furniture [32].

Nevertheless, the process of drying wood and use of resin binder will typically produce some VOC during production. It is known that OSB production will emanate VOCs while drying the wood strands [31]. Although the emissions vary according to the specie of wood and its growing conditions, terpenes are a prominent VOC that has generated during the OSB process [32]. Terpenes have been observed as the source of odorous nuisances for people living in the surrounding areas of OSB production sites [32]. Because of the prevalence of situations like these, it is imperative to explore other environmentally conscious options to provide a truly sustainable industry. Wood chips and other wood remnants that are typically unused in the lumber industry are now recycled in OSB and other particleboards. Because of the development of OSB, 80% of wood removed from the forest can now be processed into marketable products [31].

7.7 Conclusion

In order to maintain the livelihood of the lumber industry, it is critical to provide the necessary changes to the past and current norms to support the state of forests and woodland areas globally. However, the sustainability of this resource requires strict regulations that ensure the long-term health and biodiversity of the ecosystem. Unfortunately, sustainable practices have not always been applied in the past, nor are they universally applied around the world today. Because of this, technological advancements in wood and wood products have dominated the future of the industry. Presently, researchers around the world are advancing new wood technologies every day to create a more environmentally friendly industry. Not only have wood-composite alternatives been developed to promote the "greening" of the industry, but also wood preservatives and treatments to prevent the natural biodegradation of wood. Through thoughtful planning, residual wood scraps have also found a purpose in this new, environmentally conscious industry as coproducts to be recycled, and fuel to contribute back into the lumber production process. The collaborative effort from previous technologies to present-day products have mitigated the use of conventional wood for construction with more sustainable forms of enhanced wood products.

References

[1] Abbott AR, Whale LRJ. An overview of the use of glued laminated timber (Glulam) in the UK. Constr Build Mater. 1987;1:104–10.

[2] Food and Agriculture Organization. Forest area (% of land area) | Data. Available at: https://data.worldbank.org/indicator/AG.LND.FRST.ZS. Accessed: 19 July 2018.

[3] Skog KE, Wegner TH, Bilek EM, Michler CH. Desirable properties of wood for sustainable development in the twenty-first century. Ann For Sci. 2015;72:671–8.
[4] Bergman RD, Bowe SA. Environmental impact of manufacturing softwood lumber in Northeastern and North Central United States. Wood Fiber Sci. 2010;42:67–78.
[5] Bergman RD, Bowe SA. Environmental impact of producing hardwood lumber using life-cycle inventory. Wood Fiber Sci. 2008;40:448–58.
[6] Glass SV, Zelinka SL. Moisture relations and physical properties of wood. In: Wood handbook: wood as an engineering material. Madison, WI: United States Department of Agriculture, Forest Service, Forest Products Laboratory, 2010:80–98.
[7] Wiedenhoeft A. Structure and function of wood. In: Wood handbook: wood as an engineering material. Madison, WI: United States Department of Agriculture, Forest Service, Forest Products Laboratory, 2010:62–79.
[8] Czinner RL. Apparatus and method for processing lumber. 4805679, 1989.
[9] Kretschmann DE. Mechanical properties of wood. In: Wood handbook: wood as an engineering material. Madison, WI: United States Department of Agriculture, Forest Service, Forest Products Laboratory, 2010:100–45.
[10] Wiemann MC. Characteristics and availability of commercially important wood. In: Wood handbook: wood as an engineering material. Madison, WI: United States Department of Agriculture, Forest Service, Forest Products Laboratory, 2010:16–60.
[11] Falk RH. Wood as a sustainable building material. In: Wood handbook: wood as an engineering material. Madison, WI: United States Department of Agriculture, Forest Service, Forest Products Laboratory, 2010:10–15.
[12] Gustavsson L, Madlener R, Hoen HF, Jungmeier G, Karjalainen T, Klöhn S, et al. The role of wood material for greenhouse gas mitigation. Mitig Adapt Strateg Glob Chang. 2006;11:1097–127.
[13] Timber Plus Toolbox. Selecting timber: characteristics of timber, Structure of hardwoods. Available at: http://www.workspacetraining.com.au/timberplustoolbox/toolbox13_05/unit9_selecting_timber/section2_characteristics/lesson2_hardwood.htm. Accessed: 19 July 2018.
[14] Timber Plus Toolbox. Selecting timber: characteristics of timber, Structure of Softwoods. Available at: http://tle.westone.wa.gov.au/content/file/2767378d-ac34-486e-9608-4b5d0f0a4820/1/13_05_lo9.zip/unit9_selecting_timber/section2_characteristics/lesson3_softwood.htm. Accessed: 19 July 2018.
[15] Hayes D. Chemistry and properties of woods. Available at: http://www.carbolea.ul.ie/wood.php. Accessed: 19 July 2018.
[16] Grabber JH, Ralph J, Hatfield RD, Quideau S. *P*-hydroxyphenyl, guaiacyl, and syringyl lignins have similar inhibitory effects on wall degradability. J Agric Food Chem. 1997;45:2530–2.
[17] Daniel G. Microview of wood under degradation by bacteria and fungi. Wood Deterior Preserv. 2003;845:34–72. No. ACS Symposium Series.
[18] Enamul Hoque M, Maroof Khan A, Saba N, Islam MS, Asim M, Jawaid M, et al. Effects of natural degradation on the mechanical and morphological properties of tropical woods. Cellulose Chemistry and Technology. 2016;50:723–30.
[19] Singh AP, Kim YS, Singh T. Bacterial degradation of wood bacterial degradation of wood. Secondary Xylem BiologySecondary, 1 ed. United States of America: Elsevier Science Publishing Inc., 2016:169–90.
[20] Plomion C, Leprovost G, Stokes A. Wood formation in trees. Plant Physiol. 2001;127:1513–23.
[21] Kartal SN, Hwang WJ, Shinoda K, Imamura Y. Laboratory evaluation of boron-containing quaternary ammonia compound, Didecyl Dimethyl Ammonium Tetrafluoroborate (DBF) for control of decay and termite attack and fungal staining of wood. Holz als Roh - und Werkst. 2006;64:62–7.

[22] Kentucky Pesticide Education Program. Wood preservation wood preservative pesticides oil-borne preservatives, University of Kentucky, 2016.
[23] United States Envrionmental Protection Agency. Overview of wood preservative chemicals. Available at: https://www.epa.gov/ingredients-used-pesticide-products/overview-wood-preservative-chemicals. Accessed: 19 July 2018.
[24] United States Environmental Protection Agency. Pentachlorophenol, 1992.
[25] Scott J. Preservative-treated wood and alternative products in the forest service.
[26] Micaels-Glaeser JA, Lloyd JD, Woods TL. Efficacy of Didecyl Dimethyl Ammonium Chloride (DDAC), Disodium Octaborate Tetrahydrate (DOT), and Chlorothalonil (CTL) against common mold fungi, 2004.
[27] Terzi E, Taşçioĝlu C, Kartal SN, Yoshimura T. Termite resistance of solid wood and plywood treated with quaternary ammonia compounds and common fire retardants. Int Biodeterior Biodegrad. 2011;65:565–8.
[28] Vlosky RP, Smith PM, Blankenhorn PR, Haas MP. Laminated veneer lumber : a United States market overview. Wood Fiber Sci. 1994;26:456–66.
[29] Laufenberg TL. Parallel-laminated veneer : processing and performance research review. For Prod J. 1983;33:21–8.
[30] Kuzman MK, Oblak L, Vratuša S. Glued laminated timber in architecture. Drv Ind. 2010;61:197–204.
[31] Zerbe JI, Cai Z, Harpole GB. An evolutionary history of Oriented Strandboard (OSB), vol. 236, 2015.
[32] Benetto E, Becker M, Welfring J. Life cycle assessment of Oriented Strand Board (OSB): from process innovation to ecodesign. Environ Sci Technol. 2009;43:6003–9.

M. Katie Martin, Dale J. LeCaptain and Alice Delia

8 Determining wildland fire markers in residential structures using thermal desorption gas chromatography mass spectrometry

Abstract: Each year thousands of wildland fires blaze across the United States causing secondary ("smoke") damage to numerous businesses and personal property. Currently there are no specific industry standards or guidelines for determining wildfire combustion residues. Remediation decisions often rely on anecdotal evidence from occupants. A variety of particulate methods are used to assess surface contamination but there are few methods for evaluating organic chemical residues that encompass the wide range of chemical classes produced during wildland fires. A new method was developed employing a thermal desorption gas chromatography mass spectrometry (TD-GCMS). TD-GCMS using novel sorbent beds decreases the sample preparation substantially and enables sampling of bulk materials by off-gassing. Furthermore, the method developed is specific to wildland fire events. Results from a simulated wildland fire event are also provided.

Keywords: thermal desorption GC-MS, fire markers, wildfires

8.1 Introduction

2017 was a devastating year for wildfires. More than 9 million acres were burned nationally and the infamous Napa Valley wildfires alone cost more than $9 billion [1]. Negative health effects from exposure to particulate matter and wildfire smoke include asthma, bronchitis, chest pain, and congestion (head, sinus, and/or throat) [2]. The current methods for determining the composition of wildland fire emissions are extraction or derivatization combined with gas chromatography mass spectrometry (GCMS). Thermal desorption GCMS (TD-GCMS) is an easy, environmentally friendly way for home inspectors and air quality investigators to test homes for potential damages and health impacts from a wildfire.

Levoglucosan (1,6-Anhydro-β-D-glucose, CAS # 498-07-7) is a product of cellulose pyrolysis and is characteristic of all biomass burning [3]. It is formed through two transglycosylation steps when cellulose is heated above 300 °C [4]. During a

This article has previously been published in the journal Physical Sciences Reviews. Please cite as: Martin, M. K., LeCaptain, D. J., Delia, A. Determining wildland fire markers in residential structures using thermal desorption gas chromatography mass spectrometry *Physical Sciences Reviews* [Online] 2020, 5. DOI: 10.1515/psr-2019-0026.

https://doi.org/10.1515/9783110597783-008

wildfire event levoglucosan is emitted in large quantities; it is stable in the atmosphere and can be detected many miles from fire origins [3, 5].

Other categories of compounds that are characteristic of wildland fires include guaiacols, syringols, aromatics, phenols, terpenoids, and polycyclic aromatic hydrocarbons (PAHs) [6, 7]. Guaiacol and 4-R-guaiacol compounds are known softwood burning markers that are emitted in large quantities during combustion [7]. Syringol and 4-R-syringol compounds are known hardwood burning markers that are emitted in large quantities during combustion [7]. Both guaiacols and syringols are known to be present in the gas phase as well as the particle phase of combustion and there are few sources of these compounds in indoor air, therefore they would make good primary fire markers [7]. Three wildfire markers in addition to levoglucosan this study examined were Vanillin (4-Hydroxy-3-methoxybenzaldehyde, CAS # 121-33-5), Maltol (3-Hydroxy-2-methyl-4-pyrone, CAS # 118-71-8), and Syringol (2,6-Dimethoxyphenol, CAS # 91-10-1).

The most common laboratory analysis method for chemical wildland fire damage is solvent extraction for semi-volatile organic compounds such as polycyclic aromatic hydrocarbons. However, it is labor intensive, uses harsh chemicals, and does not account for other classes of chemicals or particulate matter from wildland fires.

The thermal desorption technique used for this set of experiments was off-gassing, two-stage thermal desorption. This method allows for sampling of different solid materials. Inert gas flows over a solid material while being heated; the outflow is collected onto a thermal desorption tube (TDT) [8]. The TDT is a glass tube packed with one to three different types of sorbent designed to collect specific analytes of interest [9]. When the sample is ready to be analyzed it is thermally desorbed onto a focusing trap [9]. The focusing trap is then quickly heated and the sample is injected onto the GC column [9].

Presented here is a TD-GCMS method that optimizes the recovery of the wildfire markers and proven validity by exposing residential structure materials to a simulated wildland fire event and testing the materials for the wildfire markers.

8.2 Materials and methods

8.2.1 Chemicals and materials

Levoglucosan (Sigma Aldrich, 99 % pure), Vanillin (Sigma Aldrich, ≥97 % pure), Syringol (Sigma Aldrich, 99 % pure), Maltol (Sigma Aldrich, 99 % pure), and EPA Method 525 Internal Standard (AccuStandard, 0.5 mg/mL in acetone) were purchased (Figure 8.1). Sorbents Carbopack™ B, Carbopack™ Y, and Carbopack™ F were generously supplied by Supelco®. Laminate flooring, vinyl flooring, nylon carpet, olefin carpet, upholstery fabric, and painted drywall samples were obtained.

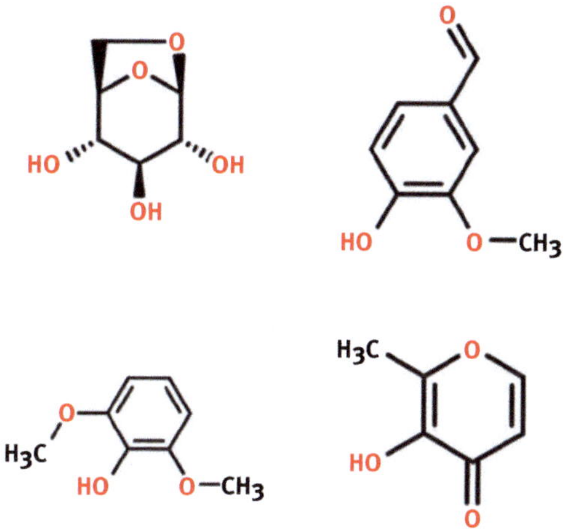

Figure 8.1: Compounds used for fire marker analysis. Top left: Levoglucosan, top right: Vanillin, bottom left: Syringol, bottom right: Maltol.

Various fresh woodchips were fresh cut (within 3 weeks of experiment) and locally sourced from Kindy Tree Services, Midland, Michigan.

8.2.2 Preparation of standards

TDTs were packed with Carbopack™ B, Carbopack™ Y, and Carbopack™ F in equal amounts. Calibration standards composed of Levoglucosan, Vanillin, Maltol, and Syringol were made at concentrations of 10, 50, 100, 250, and 500 ng/µL. Each calibration standard also had the EPA Method 525 Internal Standard added at a concentration of 100 ng/µL. TDT tubes were spiked with 1 µL of the calibration standards using the Supelco® ATIS™ chamber heated to 80 °C for 10 min.

8.2.3 Smoke chamber setup

A Thermolyne 21100 tube furnace with quartz tube was used to burn wood samples. The quartz tube was fit with a two-hole rubber stopper; one hole for air intake (air flow was 1.507 L/min into the tube), and one hole for exhaust (airflow was 1.391 L/min out of the tube). The exhaust tube was connected to a sealed drying oven. The drying oven contained all of the sample materials cut into 4 × 1 cm pieces. Samples were backed with aluminum foil so only the surface was exposed to effluent.

Equal amounts of hard and soft woodchips were added to the quartz tube. The tube furnace was heated to 750 °C at an average rate of 15 °C/min. After the wood was burned to ash the house air was directed into the oven at a rate of 20 L/min. The volume of the oven was 34.4 L. Using the oven volume and flow rate the number of air changes can be calculated; knowing the number of air changes allowed for samples to be pulled at a representative time of 0, 1, 15, and 30 days. Set-up is shown in Figure 8.2. Triplicates of the materials were removed and off-gassed after 0, 96, 1440, and 2880 air changes.

Figure 8.2: Smoke chamber setup. The outflow of the tube furnace (where the biomass is combusted) is directed into the smoke chamber (where residential materials are exposed).

8.2.4 Bulk materials off-gassing

The material was added to the Supelco® ATIS™ extraction glassware and placed in the heating block of the ATIS™. A TDT was attached to the end of the extraction glassware. The heating block was set to 80 °C; the sample was heated and purged with nitrogen for 10 min. Samples were then dry purged with nitrogen at room temperature for 20 min to remove excess moisture in the sample.

8.2.5 Instrumentation and chromatographic conditions

All TDT samples were desorbed with a Markes Ultra/Unity 2 Series TD Autosampler. The TDT was desorbed at 300 °C for 10 min onto a Markes Unity 2 general purpose cold trap. Tube split flow was 35 mL/min. The cold trap was then quickly desorbed at 300 °C for 3 min onto the GC column. The trap split flow was 20 mL/min.

Chromatographic analyses were done with a Thermo Trace GC Ultra. Injection split ratio (total split) was 49.5:1 and injector temperature was 250 °C. The analytes were separated with a TR-5 ms column (30 m × 0.25 mm ID × 0.25 µm df). The oven program started at 70 °C, hold for 5 min, ramp at 10 °C/min to 300 °C, and hold for 5 min. The MS transfer line was 300 °C. Mass spectrometric analyses were done with a Thermo DSQII single quadrupole. Ionization method was electron impact. Source temperature was 275 °C.

Data were analyzed using Xcalibur software. SIM was used to measure Levoglucosan, Vanillin, Maltol, and Syringol in samples. Precursor ions for

Levoglucosan were m/z 60 and 73, for Vanillin were m/z 151 and 152, for Maltol were m/z 126 and 71, and for Syringol were m/z 154 and 139.

8.3 Results and discussion

The TD-GCMS was calibrated and all fire markers of interest had an R^2 values of 0.99 or greater (Figure 8.3). The Maltol was only detectable at two concentrations, this is most likely due to the large split ratio. The high split ratio is necessary for a consistent sample recovery.

Triplicates of each material were pulled from the smoke chamber after 96, 1440, and 2880 air changes, which are equivalent to 1, 15, and 30 days at 4 air changes per hour. All of the samples run were very heavy, making accurate integration of peak area difficult. Maltol, Syringol, and Vanillin were recovered from all the samples (Figure 8.4). Maltol was present at concentrations above the calibration curve range even after 2880 air changes, with the exception of the olefin carpet and drywall sample which was within the calibration range after 2880 air changes The Syringol and Vanillin have a general decreasing trend within the calibration and

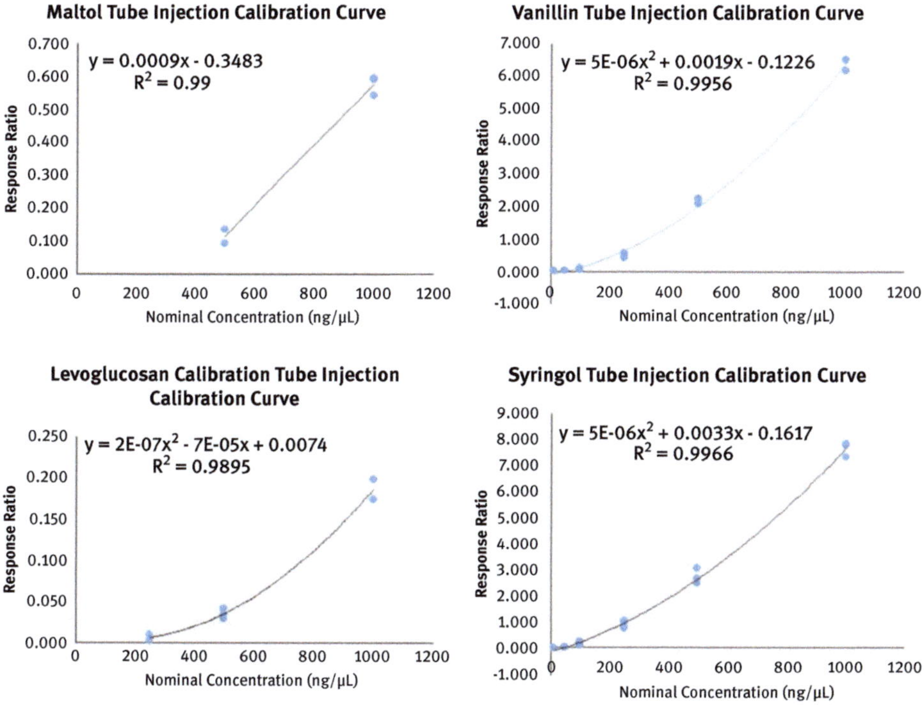

Figure 8.3: Calibration curves of primary fire markers on the Thermo Trace GC Ultra DSQ II.

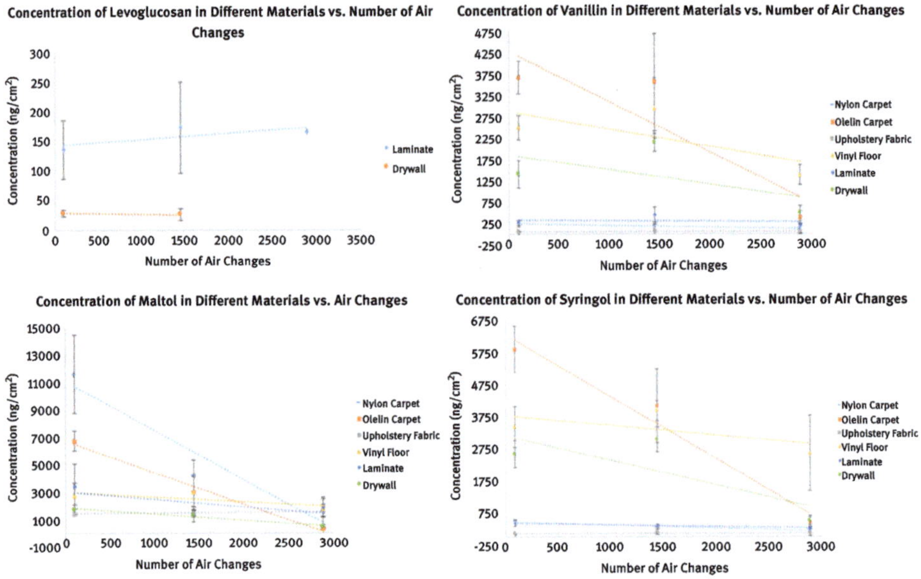

Figure 8.4: Concentrations of primary wildfire markers in various home materials after air exchanges. Top left: Levoglugosan, top right: Vanillin, bottom left: Maltol, bottom right: Syringol.

quantitation ranges. Levoglucosan was recovered from laminate and drywall samples. The laminate samples contained Levoglucosan after 96, 1440, and 2880 air changes with a large standard deviation at the 96 and 1440 points thus a definitive conclusion cannot be made about the trend. The Levoglucosan in the drywall sample was detected after 96 and 1440 air changes but not 2880. The samples show a general decrease in Levoglucosan concentration between points (Figure 8.4). Limits of detection and concentration are in Table 8.1.

Table 8.1: Limits of detection and quantification of primary fire markers on the Thermo Trace GC Ultra DSQ II.

	MALTOL	SYRINGOL	VANILLIN	LEVOGLUCOSAN
LOD (ng/cm^3)	96	16	26	16
LOQ (ng/cm^3)	320	45	85	54

Over the period of 2880 air changes the concentration of Maltol in the carpet samples had a more drastic change than the vinyl, laminate, drywall, and fabric samples. The same trend is seen in the Syringol; all samples start out with a similar concentration (except upholstery fabric) of the compound but the carpet samples have a greater change in concentration. Vanillin concentrations all start the same and remain fairly stable throughout the air changes. Levoglucosan was detected in low concentrations in only two of the materials, the error in the laminate sample was also large. In summary, Vanillin and Maltol would make good primary wildfire markers, Vanillin because of its stability over time and Maltol because of its high concentration in samples, while Syringol would make a better secondary marker. Levoglucosan would also make a good primary marker because of its uniqueness to wildland fire events. In general, the residential structural materials had a decreasing trend in the wildfire markers over time.

In addition to the wildfire markers several other fire compounds were detected. The current Prism Analytical Technologies Inc. fire marker calibration standard was spiked onto a TDT tube, run through the Thermo® system and used to identify these markers in the material samples. Compounds from the current fire marker list that appeared in the samples included 2-Furaldehyde, 2-methoxy-Phenol, 2-methoxy-4-methyl-Phenol, and p-Ethylguaiacol. Additional wildfire markers, Retene, Phenol, Syringaldehyde, and Vanillic Acid, were also detected in the samples.

8.4 Conclusions

Utilizing a TDT for wildfire markers composed of Carpobacks™ B, Y, and F for off-gassing residential home materials for the analysis of four primary wildfire markers drastically reduces the solvent and energy consumption of traditional household material smoke damage contamination analysis. A TD-GCMS method was developed for the optimal recovery of these compounds. Common residential home materials were exposed to wildfire smoke using a smoke chamber setup; samples were taken at the equivalent of 1, 15, and 30 days after exposure. Maltol, Syringol, and Vanillin were recovered from all samples while levoglucosan was only detected on laminate and drywall samples. Secondary fire markers common to all types of fire were also detected.

References

[1] Insurance Information Institute. I. facts + statistics: wildfires. https://www.iii.org/fact-statistic/facts-statistics-wildfires-Wildfires. Accessed: 12 Dec 2017.

[2] Kochi I, Loomis J, Champ P, Donovan G. Health and economic impact of wildfires: literature review and impact assessment. In: González-Cabán A, editor(s). III International symposium on fire economics, planning and policy: common problems and approaches. Carolina, Puerto Rico: DIANE Publishing Co., 2008.

[3] Simoneit BR. Biomass burning — a review of organic tracers for smoke from incomplete combustion. Appl Geochemi. 2002;17:129–62.

[4] Zhang X, Yang W, Dong C. Levoglucosan formation mechanisms during cellulose pyrolysis. J Anal Appl Pyrolysis. 2013;104:19–27.

[5] Simoneit BR, Schauer JJ, Nolte CG, Oros DR, Elias VO, Fraser MP, et al. Levoglucosan, a tracer for cellulose in biomass burning and atmospheric particles. Atmos Environ. 1999;33:173–82.
[6] Fine PM, Cass GR, Simoneit BR. Chemical characterization of fine particle emissions from fireplace combustion of woods grown in the Northeastern United States. Environ Sci Technol. 2001;35:2665–75.
[7] Schauer JJ, Kleeman MJ, Cass GR, Simoneit BR. Measurement of emissions from air pollution sources. 3. C1–C29 organic compounds from fireplace combustion of wood. Environ Sci Technol. 2001;35:1716–28.
[8] Schultz K. Capabilities of the adosrbent tube injector system (ATIS). Rep. 2013;31.2:5–7.
[9] Woolfenden E. Sorbent-based sampling methods for volatile and semi-volatile organic compounds in air: part 1: sorbent-based air monitoring options. J Chromatogr A. 2010;1217:2674–84.

Index

https://doi.org/10.1515/9783110597783-009

Bei Fragen zur Produktsicherheit wenden Sie sich bitte an:
If you have any questions regarding product safety,
please contact:

Walter de Gruyter GmbH
Genthiner Straße 13
10785 Berlin
productsafety@degruyterbrill.com